MINICAB!

MINICAB!

The True Story of Michael Gotla and the Welbeck Motors Renault Dauphine Minicabs

STEPHEN DINE

MINICAB!
is an Empress Publishing book

Published in Great Britain by
Empress Publishing
St Margaret's Road
St Leonards-on-sea
East Sussex
TN37 6EH

This edition published 2020

978-0-95641-192-1

1 3 5 7 9 10 8 6 4 2

Typeset in 13/18 pt Goudy by Falcon Oast Graphic Art Ltd, www.falcon.uk.com
Printed and bound in Great Britain

To find out more on other titles available from the author please visit
www.stephendinebooks.com

Contents

Foreword

I have not failed. I have just found 10,000 ways that don't work.

THOMAS A EDISON

Like most of us Michael Fram Gotla was a mixture of good and not so good. Educated at St.Pauls and trained in advertising, Michael enjoyed a comfortable Marylebone lifestyle. A gifted publicist with an engaging personality he was an enthusiastic entrepreneur whose attempt to modernise the London taxi trade failed because of his poor administrative control and the dismal lack of courage of his financial backers. Whilst his idea is almost the norm today, the technology of the 1960's was inadequate and unreliable and though the minicab concept was initially enthusiastically welcomed by Londoners, Michael's failure reliably to meet the demand risked the project's collapse. The subsequent pillage of the assets of the Company by image conscious financiers guaranteed its demise. It was indeed a triumph and a tragedy.

Graham Walker
London, April 2020

Introduction

On 19 June 1961, West London-based Welbeck Motors ventured from their traditional business of vehicle sales, rental and self-drive hire into what was an emerging new market in operating minicabs. They could never have foreseen fully what would lie ahead in the following months. Amidst much publicity, this brave, large scale venture took to the streets of the capital, but faced one large opponent, the London Hackney Carriage trade, more commonly known as the London black cab.

This almost forgotten story has hitherto never been told in a true, and hopefully unbiased manner and I hope to give you, the reader, the chance to discover through these chapters, a revealing tale of what is essentially a piece of history, which has changed the way in which many people now travel.

The brains behind Welbeck Minicabs was Michael Gotla, who first started in business just after the Second World War as a chauffeur hire driver, with a small developing fleet of cars. Along with his later business partner, Mr R S Walker, he gradually transformed the motor trade business, which eventually gave them a hankering to diversify into the new world of minicabs. The operation was planned in great detail and incorporated many novel ideas, such as, for example, fares of just one shilling per mile and the use of a brand new fleet of French Renault Dauphine cars.

The partners sought the financial backing of a respected well-known millionaire in order to fund the huge costs of the venture and they made use of two-way radios in the minicabs on a scale hitherto never seen in the capital; this latter was the key element of the operation's potential success. But they were unprepared for what was to come and eventually various circumstances, some beyond their control, dictated that the operation would be doomed to fail.

By all accounts, Michael Gotla was a larger-than-life character, full of energy, flamboyant and an ideas man, but he has been painted the villain by many a writer, due to the fact that he took on the established London taxi trade for a slice of the lucrative pre-booked hire business, which, at this time, they considered to be their own.

Over the years, many a story has been written, comparing the governing structures, within which the taxi trade has been obliged to operate, to the minicab trade's growth, which was free of such stringent regulations. My purpose is to record here for the first time and with the help of Welbeck Motors' own internal archives, the true hitherto untold story from this company's point of view and I hope that the reader will realise that there was, indeed, another side to this tale. As the old saying goes 'there are two sides to every story' and this is Michael Gotla's.

Preface

My interest in Welbeck Motors first started when I wrote a short article for the Renault Owners' Club's magazine about that company's fleet of Renault Dauphine minicabs. Matters started to change when I realised that, in fact, the few pieces of published information which I could find on Welbeck Motors were purely about its minicab venture and many of these small pieces appeared to be a close variation on a theme.

Eventually I discovered that, from the time Michael Gotla first established his business at the end of the Second World War up to the start of the 1960s, it had risen to become one of the largest car hire organisations in Europe and its success in new and used car sales had eventually won it a lucrative Ford franchise. After approaching businessman Isaac Wolfson of Great Universal Stores for a large loan from his General Guarantee Corporation, the Welbeck Motors business grasped the opportunity to explore new areas of expansion, which was how the idea of launching their own minicab operation onto the streets of the capital was born and became a reality.

In the late 1930's before entering the motor trade, Michael had studied law and he had discovered an interpretation within the 1869 Metropolitan Public Carriage Act, which could potentially give his new minicab undertaking the flexibility to operate outside the heavily regulated way in which the London taxi trade was obliged to operate and he decided to test its validity. Understandably, this manoeuvre infuriated the London taxi drivers, who, basically, declared war on the minicab venture. Undeterred, and by now committed with finances in place from his backers, Michael expanded this bold move, with the consequent, huge and unexpected publicity. The venture became one of the biggest national news stories of 1961.

These days, the word *entrepreneur* is overused to describe almost anyone with a business idea. But Michael Gotla had some truly groundbreaking ideas, which were greatly ahead of his time, and he had the vision and the drive to see his ideas through to the end, despite the setbacks he faced. He really was a true entrepreneur. I have been very fortunate to meet a number of former employees of both Welbeck Motors and of Michael's subsequent venture, Hill Barn Garage, and those people really knew Michael Gotla, the man. All of them quite freely referred to him as 'Mike' and all of them spoke warmly of a lively, fair, demanding, generous, creative and unique boss, perhaps even friend, who, in turn, left a mark on their own lives.

He was regularly described as being ahead of his time, of having ideas of genius and for thinking up schemes, which would take teams of PR consultants nowadays twice as long to brainstorm. He could dream them up alone and could put them into action without further help. Regretfully, I never had the privilege of meeting the man himself, so, much as I would like to have used his name Mike throughout these pages in true friendly terms, I feel that this would be too intrusive, although I hope that the reader will think of him in such a way whilst reading his story.

Acknowledgements

Early on, whilst researching the story of Welbeck Motors, I discovered that this task would not be simple as the relevant events had taken place over sixty years ago. Moreover, being able to make contact with persons still with living memories of the events would not be easy.

I am very much indebted to Mr Graham Walker, who worked for Welbeck Motors from 1957 until 1962 and whose father was the previously mentioned Mr R S Walker, the financial brains behind Welbeck. He has retained a treasure trove of information about the company, which has provided much needed accuracy to the story and without which a full and true history of the enterprise could not have been revealed.

My thanks are also due to Mr Hugh Ruding Bryan, who worked for Welbeck Motors after completing his National Service. He started out with Welbeck Motors under Michael's wing and progressed to the role of motor salesman for the organisation. Both these men knew Michael Gotla well and their first-hand experiences have been invaluable in constructing the human side of the story of Welbeck Motors. Without their reminiscences, much factual information would have been lost forever.

After Michael's time at Welbeck Motors he purchased the business of Hill Barn Garage in Sompting, West Sussex. Here my thanks must go to Roz Ford, Michael's secretary at Hill Barn, and to Tony Brooks, Graham Hawkins and Roger Ashburner, all former mechanics there in the 1960s, for their stories about Michael.

I am a member of both the Renault Frères and Renault Owners Club, and it is to the latter's former editor and regular contributor, Hector Mackenzie-Wintle, that I owe a huge debt of thanks for his knowledge of Renault's role in the Welbeck Motors saga and for his unfailing support, knowing, as an author himself, that a publication would do justice to this forgotten story. He has spent much time in reading through my manuscript, offering suggestions and editing to ensure accuracy, it has been very much appreciated.

I would like to thank author, and former London taxi driver, Bill Munro who, knowing the story from the taxi driver's side, has been a gentleman in assisting me with information, advice on my manuscript and photographs from the Stanley Roth collection. Thanks also to Graham Waite, archivist for The London Vintage Taxi Association, who has previously researched and written about Welbeck Motors.

Thank you to Andrew and Peter Gotla, Michael's nephews, for their reminiscences of their uncle in the motor trade and to Joel Frorath, son of Michael's former business partner Robin Frorath, for his memories.

Also, I put out many appeals in my quest for information in newspapers and magazines and would like to thank those who took the time to reply and to offer memories, snippets and support, including Chris Flippard, Marian Freeland, Mark Cooper, Mervyn J Thomas, Brian Gooding, Eileen Colwell, The Argus Newspaper,

John Stedman, Roger Townsend, Henrik Stenholt and to Cliff Brooker for the colourful and eye-catching jacket design, Graham How for assistance and advice and Colyn Allsopp for designing the interior.

I would like to thank the BT archive in High Holborn, London for use of their excellent facility; University College, London for assisting in finding records of Michael Gotla's time there in the 1930s; to the National Records office for freely duplicating and sending me the material which it has on file covering Welbeck Motors' brief venture into the minicab world in the early 1960s and to Richard Howes of the Pye History Trust.

Apologies if I have forgotten any further names along the way, but the assistance, which everyone has given me, has been very much appreciated.

My final and biggest thank you is to my wife Jayne and children Joshua and Bethany, who have supported me in the years it has taken of researching Michael Gotla's story and the many different paths it has taken me down!

Stephen Dine
Westfield, Hastings
October 2020

Chapter 1

The Early Years

Michael's early days and how he took a new,
but small, business venture to great heights

Michael Fram Gotla was born on 13 April 1919; his birth is recorded in the district of St George, Hanover Square, London, in the sub district of Belgravia. His parents were Framjee Shapurjee Gotla and Adelaide Teresa Gotla (née Daniels). Michael's father was a respected doctor of Indian descent, and Michael's mother had originally come from Ireland.

Little is known of Michael's early years, although, after finishing his education at St Paul's School, he registered at University College London in October 1936, in order to read law, having passed his matriculation exam in January of the same year. He left in the 1937–38 session and went into advertising as a copywriter for the firm of Mather & Crowther Ltd, although he had hankerings for the car business. As Britain entered the war years, Michael served in the RASC (Royal Army Service Corps) and was eventually invalided out; he then went to work for Hawker Aircraft and after that for the Ministry of Information.

In the post-war period, Michael returned to his original employment, (Mather & Crowther Ltd) but, with his keen interest in motor cars, he soon established Welbeck Motors with David Evans, an ex-Merchant Navy man. On a capital of fifty pounds and using a car on hire purchase, they did chauffeuring, with David driving during the day and Michael mostly at night. Michael was quoted as saying 'My father was one of those people, who thought that the motor industry was the end' and initially things were done very much on a shoestring; apparently both men even shared the one chauffeur's cap! But, with hard work, they bought two more cars and, being based in small premises at 107 Crawford Street, London W1, they were conveniently situated not far from Baker Street station.

The new business' first listing was in the May 1946 edition of the telephone directory; telephone number WELbeck 3991. On 5 December 1947, Welbeck Motors was incorporated as a Limited, Liability company, with the two subscribers being listed as Michael Gotla, living at 87 Gloucester Place, London W1 and Alan Edward Kerhoe of 22 Pennant Mews, London W8, both being motor car hirers and both with one share of £1 value. 'The latters' name will re-appear later in the story as a director of Welbeck Motors (Car Rentals) Ltd. Around 1949 a switch was made from a chauffeur-driven to a self-drive car hire business, supported with money borrowed from a finance company. By then, David Evans had sold out, on condition that Michael paid the then outstanding crippling £200 telephone bill, and he was said to have gone back to sea. This showroom itself could display only two cars inside, but, eventually, the adjacent site was taken over and plans were made for a new, purpose-built showroom, which in due course would become a reality.

A contract to supply American GIs in the UK with rental cars via the American Forces helped to establish the business, which eventually consisted of about twenty cars, thought to be Austin A40s. In the post-war years, it was not easy to obtain a new car for the home market, but because of this GI connection, Michael's business was considered to be supplying the export market for visitors to England and thus he had been able to obtain a special license to purchase new cars direct from the factory. After the cars

Smog in Crawford Street, London circa 1940s. This atmospheric photograph was taken outside the pharmacy of Meacher, Higgins & Thomas (Established in 1814). Welbeck Motors hire cars can be seen parked in the street. (Mervyn Thomas)

had been hired out for the first year, they could be sold legally and in fact, were sold for more than they would have cost when new! It is worth recalling that, at the time, there was often a two year waiting list for such vehicles!

In February 1951, Welbeck Motors was still listed as being at 107 Crawford Street, but the company also had premises at 75 Baker Street (Tel: WELbeck 1139), although the entry for this address does not appear again in the next telephone directory.

Michael had a brief taste of what it was like to be on the front pages of the national papers, albeit only in the small print. In the June national papers, it was revealed that two missing British diplomats had recently hired a car from his company. On May 25, one of the diplomats, a certain Mr Guy Burgess, went to Welbeck Motors to hire a new cream Austin A70, registration VMF 196. Having ordered it from the Reform Club's phone, Burgess called round at 2pm in Crawford Street and paid a £25 deposit, and drove off to collect fellow diplomat, a certain Mr Donald Maclean.

Michael stated that Foreign Office officials often hired cars from his company. Armed with an RAC map of southern England showing them the route from London to Southampton, with a street plan of London and with a plan of Southampton Docks, the diplomatic duo made a dash for Southampton, arriving just before midnight. They were witnessed by workers roaring into No 9 docks', nearly hitting a truck in the process, before pulling up and abandoning the car. One of the men threw some money at one of the dock workers, saying, 'Buy yourself a drink', before just managing to board a departing ship. The car was put in a parking bay by the workers and then left there. Eventually, as the renegade diplomats' full situation came to light, Michael read of the men's suspicious disappearance from the country in the papers, and, realising it was one of his cars, which had disappeared with them, he called the police. The car was searched and Welbeck Motors, who had been looking for the car for a fortnight, were contacted to confirm the car's details. The two men's story continued thus. After eventually docking at St Malo in France, they took a taxi to Rennes, before catching a train to Paris. The British embassy in Rome had asked the Italian police for help in the bizarre case of the missing men, at least one of whom it was thought at the time might have travelled there. And the rest is history for the defecting diplomats . . .

Car sales, both new and used, would become a large part of Welbeck Motors' operation, first being listed as such in the April 1952 edition of the telephone directory, still with the original contact number, although it is possible Michael was testing out this new market before changing direction. Advertising at the time billed them as 'London's most progressive car sales organisation' soon a newer number WELbeck 1139 (6 lines) would be in use.

In one of Michael's many pieces of advertising that he wrote for inclusion in trade press magazines and newspapers, a full-page advertisement in 'The Autocar – London Show Review Magazine' dated 26 October 1956, had been taken out and was entitled 'Welbeck Motors Ltd. The First Three Years'. His story begins with how he had changed his mind in the October of 1953 from wanting to run just car-hire, although it was a fine business, it was steady but dull. Almost overnight Welbeck Motors had 'started anew' and were not just going to buy and sell cars – they must become outstanding in this field: Welbeck Motors must have a famous name in car sales – *quickly*.

Michael wanted a famous manufacturers name to be associated with Welbeck's. He knew it would not be easy starting in this new area of dedicated sales, especially in the heart of London. He initially signed with a Jowett car agency, with his staff being trained in sales and service for the product, but it was with much regret when the Jowett company soon after had had to make the difficult decision to cease production due to what had been a steady fall in sales.

Michael did secure a second agency with Armstrong Siddeley, the feeling that both the company and their London distributors were kind and helpful to work with, although regretfully this arrangement lasted just over a year. With no ready market in Michael's eyes readily available for the supply of new cars, he decided to focus on obtaining nearly new stock cars of this type, with every car having to be sought out and every customer found, in order to buy and sell to slowly increase sales.

Bulk buying of stocks of cars had brought success to the business and a number of new but unused 18 hp Armstrong Siddeley cars, after this model had ceased production in 1954, had been successfully purchased and sold on, followed by Jowett Jupiter's from the Bradford based company.

A busy Crawford Street in 1954. Some of the Swallow Dorettis for sale outside the original Welbeck showroom. (Hugh Ruding Bryan)

On occasions, Michael secured deals on batches of interesting cars coming up for sale, which would be considered quite collectable nowadays. Salesman Hugh Ruding Bryan recalled being sent by 'his boss' to collect the remaining eleven Swallow Doretti two-seater sports cars, when the factory went out of business. Built in Walsall, Staffordshire of tubular chassis construction and based on Triumph TR2 mechanics, they had been produced between 1954 and 1955. Within this selection of cars, he recalled that there was one prototype four-seater, with black leather seats and wire wheels which Michael had said he was worried about being able to sell as a one-off car. However, it did sell, but for less than the others, at £645. Michael later stated that taking on the last of these Swallow's had created 'National interest'.

Another new car, which that was being marketed by Welbeck Motors in August 1954 was the Morgan sports car. Their advertising billed it as 'Proudly introducing the wonderful new Morgan'. Fitted with a Triumph TR2 engine, it was claimed that, at a cost of £830 inclusive, it was the lowest-priced fully-equipped over 100mph model on the market. However, it was noted by the salesmen at Welbeck Motors that this marque did not sell particularly well at the time, so not many examples passed through the showroom doors.

Architects plans were drawn up in March 1955 for a new, purpose-built showroom next to the Crawford Street site. A large block of bomb damaged property had been acquired and demolished with plans for the new facility to have a frontage of (approximately) fifty-five feet and a depth of forty-two feet, a basement for an underground car park, complete with a lift, and a ground floor level with a mezzanine for a suite of offices for the business, plus four flats on the two floors above. It would be constructed of reinforced concrete with brick filling to the external walls and the front elevation would have Portland stone dressing. The small original car-hire reception centre was turned into a temporary showroom with additional stock kept in nearby garages until the transition to the ultra-modern showroom.

By February 1956, the work was well under way, with the business finally moving in ahead of the Christmas festivities the same year. It was described as 109–111 Crawford Street.

Occasionally, other motoring rarities came into stock, such as in June 1956, when a deal advertised as 'the most sensational new car offer of all time' – it was the Paramount. Michael had bought in what was described in the advertisement as the maker's entire stock of this Leighton Buzzard-built product. It consisted of twenty-six, four seater convertibles and three saloons, all fitted with Ford engines.

WELBECK MOTORS *proudly* *introduce the wonderful new* MORGAN

The "Autocar" Road Test of May 7th focused world attention on this amazing new car. Fitted with the TR2 engine and with greatly improved appearance and handling qualities, it is (at £830 inclusive) the lowest priced, fully-equipped over-100-m.p.h. model on the market. Cars available at our showrooms for inspection and trial or write for full details and performance figures; delivery is quick and certain. THE MORGAN HAS AGAIN ENTERED THE VERY TOP GRADE OF THE WORLD'S SPORTS CARS.

THE EASIEST WAY TO WIN A RALLY IS TO BUY A MORGAN

WELBECK MOTORS, LTD - 107 CRAWFORD ST., LONDON, W.1

Near Baker Street Station - Telephone : WELBECK 1139 (6 *lines*)

★ ★ ★ **Unbelievably easy Hire Purchase Terms now available**

Advertisement for the new Morgan. (Motorsport magazine August 1954)

The maker's retail price had previously been £1,013, but Welbeck Motors were selling them for the 'value for money – beyond belief' price of £795, including tax. They were known to be badly made and the interiors were shabby. Constructed with ash frames and an aluminium chassis, nothing fitted well either. They looked great – but the build quality was poor.

Michael had made note that it was more of a gamble to buy the whole stock rather than just a few, although the cars all sold and felt this 'coup' had kept the Welbeck name in front of the public.

Michael was delighted when the Rootes Group chose Welbeck Motors to sell and service the new Singer Gazelle model, granting them the wholesale rights for what Michael considered 'an important part of London'.

The November 1956 telephone directory listed a property at 22 Crawford Street as well as at 107, which indicates that the construction work of demolishing the original premises and building the new showroom was in progress, as by the February 1958 edition, number 109 was listed. This would be the ultra-modern purpose-built premises, which would give the business the much-needed breathing space it required. 22 Crawford Street was still listed as being in use at the time of the April 1959 directory, but it does not appear thereafter.

The year 1956 brought to Welbeck Motors an important figure, who was to shape the future of the business. Mr Rowland Stuart Walker (1913-1997), known in the trade as Mr R S Walker, but, to close friends, as 'Jock', he being a Scotsman from Aberdeen. Walker had first met Michael at a trade cocktail party that year and both men had hit it off. Michael needed help with restructuring his

Photo by courtesy of "The Motor"

Welbeck Motors make the most sensational new-car offer of all time

BRAND-NEW PARAMOUNTS at £795

including Tax

As a result of a breath-taking bulk purchase, Welbeck Motors are able to offer the entire Makers' stock of Paramount Cars at £795 including tax. These craftsmen-built cars have previously been priced at £1,013 and at the new figure represent a standard of value-for-money beyond belief. The Paramount is a full four-seater of outstanding appearance: the chassis is tubular; the coachwork is of light alloy on a frame of rot-proofed seasoned ash. *No car is better made or finished with greater care.* Ford and other proprietary components make servicing cheap and easy: spares are readily available. If you want one of these cars, treat the matter as urgent: we have only 26 and at the new price they are going to sell very quickly.

WELBECK MOTORS LTD

107, CRAWFORD STREET, LONDON, W.1

(Near Baker Street Station)

Telephone: WELbeck 1139.

There's always something interesting happening at Welbeck Motors—London's most progressive Car Sales Organisation.

The following cars are available :—

21 *four-seater convertibles fitted with Ford Consul engine.*

1 *four-seater convertible fitted with Ford Ten engine.*

1 *four-seater convertible fitted with Ford Consul engine, left-hand drive. Available for export free of purchase tax.*

3 *saloons (same appearance as convertible but with hard-top) fitted with Ford Consul engine.*

Advertisement for the new Paramount. ('Autocar' 8 June 1956) financial commitments with Lombard North Central, with whom Walker had a good reputation in business. So, with his considerable financial expertise and with ideas on further expansion for Welbeck Motors, he assumed the position of chairman of the organisation, with Michael becoming managing director. Having left the army some 18 months previously, Walker had moved to Maidenhead in 1948, and he had initially bought a garage on the Cordwallis Estate, where he made tractors for a time, and he eventually expanded his business to Reading and Slough.

In 1957, a new scheme was introduced, which was called 'Fords for the Forces' where, if one was serving in the Armed Forces and were to be posted overseas in the near future, one could take delivery of a new Ford, free of purchase tax. The car could be used in England for up to six months (or twelve months if one had already served overseas in a recent period), whereafter it could be taken overseas with the posting. The deal required the vehicle to be kept abroad for a year, and then retained for a further two years after returning to the UK. In this way, the vehicle escaped the payment of purchase tax altogether. Welbeck Motors could offer hire purchase facilities

Michael Gotla (pictured right) with Robin Frorath outside the new showrooms at Crawford Street, London circa 1956 (Hugh Ruding Bryan)

and would even lend the money towards the shipping costs. A part exchange facility was still offered, and the person could even buy a car whilst abroad, have the car shipped overseas and have the financial details and documentation posted to them.

In a similar manner, visitors to Britain could also buy a car from Welbeck Motors free of purchase tax, could use the vehicle in the UK for up to twelve months and then take it home after their stay here. To add to the excitement of the deal, Michael offered a guaranteed repurchase scheme, by which Welbeck Motors would buy the car back for a pre-determined sum. This was agreed before the client was to return home and was based on the car's depreciation during the length of time the person was in the UK. Experience had shown that many visitors would be able to have the use of a new Ford for three months at a cost of £100, which was great value for money in those days.

In February 1959 No. 95 Crawford Street was leased together with nine flats above. Michael named the new acquisition 'The Volkswagen Room' and it was situated further down the road, next to what is now 'The Duke of Wellington' public house. An advertisement in 'Punch' magazine for January 1960 describes how the nickname was said to have started as a joke by a salesman who wanted to buy and sell used Volkswagens, because he felt that Welbeck Motors were too Ford minded! The brand new showroom was described as housing 'London's finest show of Volkswagen Beetles'. Welbeck Motors had applied for a Volkswagen dealership, which had been refused, but they were, as they put it, 'quite uninhibited'. Although they were official dealers for Ford, the Volkswagen Room was free of all the restrictions of a formal franchise. By this time both telephone numbers had changed to WELbeck 0561.

Welbeck Motors vehicle road fund licence
holder – circa 1950s. (Hugh Ruding Bryan)

At the time, Volkswagen 'Beetles' were very saleable cars and
this separate showroom was devoted solely to this German marque.
Noted for their great reliability, their air-cooled engines always
started after standing in the showroom for longer periods than
normal. A comparison was obvious against some of the British
makes, which had their problems. A second-hand Jowett Javelin,
for example, would always succumb to a flat battery, which would
simply discharge overnight, and, the next day, if it needed moving,
the use of the starting handle would be required to bring it to life.

As space for vehicles was always at a premium, sixteen garages at
Durweston Street, along with five flats at Durweston Court, were
also rented at £3,000 per annum from June 1960. Except for one
garage, which was owned by a certain lord, who would neither rent
nor sell it to Welbeck! One garage was used as a car wash bay, with
the others housing vehicles, either for renovation or as trade cars.

Around 1957, Welbeck Motors had taken on a Ford franchise, and,
on one notable occasion a regular delivery of cars arrived direct

from Ford at Chailey Street, Tottenham Court Road. The cars were unloaded from Welbeck Motors' own car transporter, in order to be de-waxed and to be cleaned in preparation for either customer collection or for the showroom. About an hour later, after this particular preparation task had been finished, the sales manager went over to the showroom and said 'You're not going to believe this, but you had better come and take a look at this car.' The showroom staff obliged and the manager pointed out the problem. 'Did you spot it? It has two doors on one side and one on the other!' That was exactly how the car had arrived from the factory. The body shell was the same for both two and four door versions, but there had been a mix up on the production line with what one assumes must have been a different team each building on its own side of the line, which had ensured this uniquely constructed new car had passed Fords' quality control and . . . the vehicle had actually been delivered to Welbeck Motors' showroom for sale! Understandably, Michael sent the car back to the factory.

Scottish businessman and philanthropist, Isaac Wolfson's General Guarantee Corporation was approached for financial assistance to enable Welbeck Motors to expand into other areas of business. Mr Wolfson's family trust purchased 51% of the company, buying 14,898 shares in Welbeck Motors for £14,898 on 6 April 1959, when it also injected what was reported to be a loan of £1,000,000 into Welbeck Motors, in order that the company's plans could be realised. The Corporation's involvement came about after Welbeck Motors had approached finance houses in the City to invest in the business. They were interested, but wanted what Michael described as 'an enormous slice of the cake' to do so. The Corporation on the other hand were prepared to take shares in the company and agreed that the business should carry on as before. Several shareholders were said to be part of the minicab scheme. Lombard Street Holdings, which was a subsidiary of Martins Bank, then the sixth largest bank in Britain, was one of them.

In addition to providing a new Ford agency for the Baker Street area of London, the deal also provided the General Guarantee Corporation (a hire purchase company) with a sales outlet for any vehicles, which it re-possessed. Moreover, it provided a source of hire purchase business for that division of Welbeck Motors. Michael signed an agreement to continue to act as managing director for a ten year period on a salary of £1,500 per annum with a commission of 5% per annum on the net profit of the company. Part of the agreement, drawn up on 6 April, was that Michael could continue to live rent-free in Flat 2 on the first floor of 109 – 111 Crawford Street, above the showrooms. Mr Walker was also formally offered the post of chairman on a salary of £500 per annum, the use rent-free of Flat 1, 95 Crawford Street and the use of a motor car.

The man responsible for putting this deal together Mr David Finnie of Finnie, Ross, Welch & Co, Chartered Accountants, was appointed to the board of Welbeck Motors in the June of 1959. Finnie had been an adviser to Isaac Wolfson for many years and was, at that time, chairman of the Corporation. In October he was joined by Mr R A J Emery, who ran Ralli Bros Ltd for the Corporation.

Isaac Wolfson (1897-1991) was of Jewish background and he had worked his way up through the ranks of business to become the managing director of Great Universal Stores in 1932, and from 1947 until 1987 he was the chairman thereof. He established the Wolfson Foundation in 1955 for the advancement of health, education and youth activities and, in the following years, donated millions of pounds to its causes. Wolfson was never shy of philanthropy and he is understood to have once said 'No man should have more than £100,000 – the rest should go to charity'. He was later knighted.

As part of Welbeck Motors' renewal of their car hire fleet, and with the Corporation's backing, Michael visited the headquarters of the Ford Motor Company in Regent Street. Two main subjects were discussed: a fleet order for one thousand cars for Welbeck Motors (Car Rentals) Ltd (although he anticipated receiving only about half of this order) and the purchase of a Ford main dealership somewhere within a thirty mile radius of London. This was said to be the next logical step in the company's development with the assurance from the new backers that the necessary finances would be made available. An order was placed the following year (1960) worth £430,000 for six hundred new Fords in a deal, which was done ahead of the opening of the Motor Show in October. At the same time, Michael Gotla placed an order with Renault for two hundred vehicles. At that same year's Commercial Vehicle Motor Show, Michael ordered four hundred Ford Thames light vans, at a cost of £160,000, for a new commercial vehicle leasing system, which he was yet to announce.

For Welbeck Motors, both car and van hire were buoyant as the year turned 1960. For example, the large and regular sales campaigns in the press announced that one could hire a new Thames 5cwt* van for as little as £2 19s 6d per week, depending on the contract period, or for even less for a fleet owner. Deals on the 7cwt*, 10/12cwt*, 15cwt* vans and the Thames 12-seater could also be made, the

* cwt = hundredweight

Rate Chart for
Private and Commercial Vehicles

The

Rental System

at

Welbeck Motors

WELBECK MOTORS LTD
The Welbeck Building, 109 Crawford Street, W.1
WELbeck 0561

Brochure for Welbeck Motors rental hire charges
(Graham Walker – Welbeck Archive)

advertising pointing out that 'Obviously hiring is the most trouble-free method of running a van.' Big contractors were being advised that Welbeck Motors could issue long leases on heavy lorries and motor coaches. 'May we tell you more?' was the slogan to start an intensive campaign to popularise both short- and long-term hire of commercial vehicles.

A *Night Season Ticket* 'Car hire at the price of a bus ticket' promotion was already into its second year. Initially five hundred night season tickets were offered and, if a customer paid Welbeck Motors the sum of £18 18 0d (18 guineas), they could hire a car from them for as many evenings as they wished for a year between 6pm to 9am (except Saturdays and Sundays). Thousands of applications had been received and so great was the interest in the deal that over one hundred newspapers carried special stories about it. Whole page advertisements were a regular feature of 'Autocar' magazine, where promotions on 'Three ways to run your Ford' would catch the reader's eye.

Options either to buy a car or to sign a long lease or to short term hire on the occasions when one needed a car, were being offered. For long term hire on private cars, the advertising stated that, with this dynamic Ford dealership, one hundred cars were on show for immediate delivery – Ford Anglias, Consuls and Zephyrs were available. Welbeck Motors also received an accolade from Renault for selling a record number of their Dauphines. As the company's trading activities grew, a new company was set up. Welbeck Motors (Service) Ltd was incorporated on 23 May to divide up some of the many, varied aspects of the operation.

In August, Michael reported to the board that he and the chairman had been offered an eighteen-year lease on a delicatessen shop in Crawford Street at an annual rent of £685 plus £2,850, that sum to include the stock in hand at the date of entry. It was agreed that this offer should be taken up by the company; the premises were to be developed as a snack bar as well as a delicatessen, and the staff of all three companies would be encouraged to use it as a canteen.

At this time, further possibilities were being presented to Welbeck Motors. A small car hire firm in Bradford had made inquiries with Michael about a possible sale, presenting him with an opportunity for expanding the group's car hire activities, but eventually this was not pursued. The emphasis was on expanding the number of branches in the London area, and these would stretch eventually to Ilford, Kingston, St Albans, Slough and Cheam as well as Crawford Street. Indeed, business was flourishing. The car, van

Michael Gotla with pre-delivery Ford cars. (Roz Horner collection)

and commercial rental businesses were reported to have in the region of 2,500 vehicles out on short and long term hire. Before the Corporation's financial backing, the car hire fleet stood at a more modest twenty-five cars, but this had now grown significantly to fourteen hundred. In fact, an article on contract vehicle hire, published by ERC Market Research in May 1961, suggested that Welbeck Motors' market share was the second largest in the UK, only topped by Godfrey Davis Ltd, with companies such as Roy Galway and Transport Rentals below Welbeck in the percentage league tables. Michael himself was enjoying the rewards of the fast expanding Welbeck Motors organisation; he was reported to have two personal cars, his London flat, a seaside retreat in Rustington, West Sussex, with a speedboat nearby, and he was in the process of converting an old Georgian house in St Albans, as well as a small grocery shop.

THE MOTOR   July 26 1961

WELBECK MOTORS
Thames 10/12 cwt. Van

SHORT TERM HIRE A unique service: when *your* van is off the road or *you* need a van on an odd occasion, call on Welbeck. A Thames 10/12 cwt. van costs only £1.15.0. for an evening; £2 for a day; £3 for 24 hours, or £15 for a week. For a Monday to Friday hire a special rate of £9 applies. During the winter and for longer hires even lower rates. These prices are absolutely all-inclusive, there are no extras, no mileage charge, no mileage limit and no-one cares how far you go.

LONG TERM HIRE On a two year contract a Thames 10/12 cwt. van, taxed, painted, fitted with heater, passenger seat, extra sun visor, foot rail and quarter rear bumpers, plus free repairs throughout the contract and the loan of a replacement vehicle in the case of accident or breakdown, costs only £21.11.8 per month. Should free repairs and replacement of vehicle not be included in the contract then the price reduces to £19.10.0. per month. If you prefer a one year contract, the basic rentals quoted above are increased by 25%. In the case of business users the hire charges are allowable in full by the Inland Revenue.

PURCHASE A special member of our staff deals with commercial vehicle sales. A Thames 10/12 cwt. van painted, fitted with heater, passenger seat, extra sun visor, foot rail and quarter rear bumpers costs £494.10.0. Delivery is immediate with a complete choice of colours. A far-sighted part exchange policy, low interest hire purchase and, of course, the astonishingly obliging Welbeck Service Department all make it very pleasant for the customer.

The Thames 15 cwt. van is also available at fractionally higher rates. Special discounts for fleet owners

WELBECK ⟨wm⟩ MOTORS

The Welbeck Building
109 Crawford Street, London, W.1.   Telephone Welbeck 0561.
Sales Dept. open until 7 p.m. Monday to Saturday: Hire Dept. open every day until 11 p.m.

BRANCHES FOR SHORT TERM HIRE ONLY AT:
ILFORD (GOODMAYES 3584); KINGSTON (KINGSTON 3492); SLOUGH (SLOUGH 25363); ST. ALBANS (ST. ALBANS 50615); CHEAM (VIGILANT 0695).
TELEPHONE YOUR LOCAL BRANCH FOR ADDRESS AND HOURS OF BUSINESS

A1

Welbeck Motors advertisement for hire or purchase options on the new Thames 10/12 cwt. Van. ('The Motor' 26th July 1961)

Chapter 2

Plans for a Minicab Operation

With Michael's new financial backing from the General Guarantee Corporation, he now had the realistic chance of diversifying his business interests and of trialling a new conception of transport, which, he visualised, would revolutionise how Londoners might want to travel for a reasonable fare in the future.

With the advent of a number of businesses making plans to launch minicab operations in the capital over the coming months, the established monopoly of the London hackney cab gradually came under threat at the dawn of the 1960s. If these operations were successful, the likelihood of other small ventures gaining momentum across other parts of the country was causing considerable speculation in the press. The general taxi trade opposed this new type of operation, because there was a huge difference between a car hire service and a taxi business. For example, the car hire fares would be substantially cheaper; there was the issue of how the newcomers would operate within existing regulations, to which the traditional taxi trade was obliged to adhere. There was the grey area of how the newcomers might 'ply for hire', for example, when a passing member of the public hailed a car, the hire firm was legally unable to respond. The taxi industry felt that it needed to take united action to oppose what could be a substantial loss of trade.

Another major factor was the London taxi drivers' renown for knowing the topography of the capital, known by many simply as 'The Knowledge', for which they had been obliged to study hard, and at some expense, over a period of one to two years. Drivers would have been obliged to cycle around the capital's streets for, on average, thirty miles a day and to have studied at night in order to obtain their licences. The idea that almost anyone could gain employment by jumping into a minicab, to be free to pick up fares without any real knowledge of the capital and without being regulated in the same way as was the traditional taxi trade, was not evenhanded. A taxi driver could easily be identified by his green badge, with his personal number on it; there were also yellow badge holders, who would also have had to have passed their 'knowledge' tests, but only for the area in which they were licensed to operate. However a minicab driver would have no means of identification at all, and his services would be completely unregulated, which could seriously compromise any hirer's safety. Travelling by taxi was (and still is) a big industry and, at that time, over fifteen million pounds per year was spent on the use of taxis in London. However, some members of the public felt that there were simply not enough taxis to meet peak period demand in the City.

The idea of running cabs with radios was not a novelty in itself, as, on the 1 December 1947, the Company, Cosy Cars of Streatham, London was noted to have started the first radio service in the city. It would be the advances in two-way radio technology, which would be crucial to the potential success of minicab operators' new ventures.

An ally of the new minicab ventures was Rupert Speir (1910–1998), who had been Conservative MP for Hexham, Northumberland since 1951 and who had first mooted the need for smaller taxis in the then well-known debate in the House of Commons in November 1960. He was critical of the contemporary London taxicab service, but not of the taxi drivers themselves. Mr Speir felt that the introduction of minicabs would provide the London public with a better overall cab service and he felt that the existing service was inadequate, it was deteriorating and he opined that the system needed examination in order to encourage the possibility of fixed bookings in advance of travel, and that the time had come, as he said, 'for us to have minicabs'.

Certainly, the traditional monopoly of the hackney carriage trade could not always provide the public with the necessary service at peak times. It was reported in the papers of the time that if it rained, taxis were virtually unobtainable, and, during rush hours in Central London, a wait of ten minutes or more was quite commonplace, with potential customers waving desperately from the side of the road, attempting to hail a cab. In 1931, years before the advent of the minicabs, there were said to be in the region of 8,203 taxis on the streets of London, but, at the end of 1960, there were only 6,656. The press commented that the attitude of some taxi drivers left something to be desired and, on the face of it, the public appeared happy to embrace the minicab, this new form of transport, which would gradually become an ever more familiar sight all over the country.

An early starter in minicab operation was Carline, based in Wimbledon, Surrey, which began operations with radio-controlled vehicles on 6 March 1961, using a modest fleet of three (some reports stated twelve), two-tone blue Ford Anglias, which were reputed to have carried five hundred passengers in their first week of operation. Carline hoped to increase the fleet to fifty vehicles by the following month. The start of the operation was successful and general manager Ronald Heath was pleased with the large public interest in the new venture. On June 5, Carline went one step further by publicising 'free' journeys for passengers who arrived at London airport and at all the London mainline railway stations, and who would then be travelling in the direction to the company's home base in Wimbledon. The minicabs would wait for fifteen minutes after dropping their original fares off at such termini and then they would display a 'free lift' notice in the windscreen, which would apply, even if there were no advance bookings for a return journey.

Within days of the Carline operation starting up, Sylvester Car Hire of Dolphin Square, Millbank, which had been established in 1957 by Tom Sylvester and Major Hilary Scott, launched a minicab service in the capital on the 20 March, using a fleet of twenty five Fiat Multipla cars, which were supplied by J Davy & Co, the main Fiat distributors in West Central London. The Fiats had a distinctive, what we would now call MPV, look to their styling. Operated by owner-drivers, they ran on a telephone-based booking system, whereby the drivers, who were out on the road, would be directed to their fares via a Hudson two-way radio on board.

These cars were primarily dispatched throughout the Chelsea, Westminster, Kensington, Mayfair and City areas in order to pick up fares, who had booked by calling VICtoria 3691. Mr Sylvester (who, incidentally, was a former circus artiste and son of TV actress Ingrid Sylvester) was no stranger to this type of scheme, because, back in 1958, he had piloted an experimental scheme using standard Ford Consul saloon cars equipped with radios and meters. Demand was soon proven with the service mainly in the Westminster and Kensington areas receiving up to seven hundred calls per day; they were said to be 'rushed off their feet' with all vehicles being kept fully occupied, which amply demonstrated the demand for this new kind of service.

An example of the Fiat Multipla 600 cab version called 'Multihire', which was shown at Earls Court Motor Show in London, before delivery to Carline. ('Car Hire' magazine)

It was one of Sylvester's vehicles, which, on 24 March, became embroiled in what became known as 'the Battle of Belgrave Square' where a lone Fiat, driven by Mr Jim Buntin, became boxed in by a number of taxis, who, mistakenly, thought the driver was plying for hire. It was not long before a reported two hundred irate taxi drivers had converged on this unfortunate minicab and its driver, who had been told via radio to wait there, after dropping off a fare at Knightsbridge. An ugly scene developed and it was only after the police had arrived and had threatened the taxi drivers with prosecution for causing an obstruction, that the taxi drivers gradually dispersed. A spy camera was soon fitted to one of the Fiats, in order to collect evidence of any further such encounters with taxis. On 1 April, its fitment was justified, when a minicab driver and his vehicle became surrounded by nine taxis, while waiting in Duke Street for instructions for its next booking. Police were called and the whole procession moved off to West End Central Police Station, whilst the camera continued to roll. The film was studied to see if there had been any wrong-doing caught on film. This was to be a sobering taste of what Michael would shortly be facing, when his own operation began, due to the fierce opposition from the taxi trade to this new kind of transport in the capital.

Indeed, when Welbeck Motors' intended plans to launch their own large fleet of two hundred minicabs as a separate side to their main car rentals and sales business went public early in December 1960, the press wasted no time in printing articles on 'Cabs at war' and 'Battles between minicabs will begin next summer'. They drew attention as to how these new ventures might change the public's travelling habits.

What, then, was to make Michael's proposed venture different from other small businesses already operating minicabs? Primarily, either a potential passenger could telephone Welbeck Motors for a minicab, a system which the other firms were already operating, and which resulted in the nearest minicab being re-routed to the waiting fare. Or, unlike the other firms, the potential passenger could hail a Welbeck Minicab either when it was stationary or it was in motion, whereafter the driver of the minicab would call headquarters via his on-board, two-way radio in order to book a vehicle for the client. In turn, HQ would then radio the next nearest available minicab and despatch it to the now waiting passenger. It was argued that, by this method the first minicab had become a mobile booking office and that this constituted an advance booking, which had been made for the second vehicle. Legally, the passenger could not board the vehicle, which he had initially hailed, as this would be deemed to be 'plying for hire', which was in breach of the regulations. Michael, with his knowledge of law, had found what could be a loophole in the Metropolitan Public Carriage Act, 1869 and had decided to test out how operating in such a manner could not be deemed to be plying for hire. At the same time he felt that, if this ploy was proved not to be legal, he would revert to operating his service on a telephone booking system only.

As far as tariffs were concerned, Sylvester's Car Hire's schedule had been set at 1s 6d for the first mile and 1s thereafter, whereas Michael decided on a competitive 1s per mile, whatever the distance, with no extras and no return charge. This meant that if up to three people were carried, together with luggage, it could hardly be any more expensive for them to travel thus than by using public transport. Traditional London taxi fares of the time were 2s 3d for the first mile, which included a 1s 3d initial hiring fee, then 1s 3d thereafter up to six miles, with extra for additional passengers and with any distance over six miles being a matter of bargaining.

Understandably, this new type of service was opposed by the traditional taxi owners, who had been appealing already for the control of entry into their trade at the time. From the public's point of view, the suggestion of cheaper fares for travel was met with general approval for the start of a new operation in the capital. A

One of the protagonists of the mini-cab movement in Parliament is Mr. Rupert Spier, Conservative M.P. for Hexham. He is seen here trying one of the Sylvester cabs after inaugurating the service last month

Rupert Speir M.P. a protagonist of the Minicab movement.
('Car Hire' magazine March / April 1961)

novel idea of Michael's was that some of the operating expenses of Welbeck Minicabs would be met by selling advertising space on the sides and rear of the minicabs themselves (more details in Chapter 5), which would help to offset the competitive fares being offered. This, in itself, was a revolution, which would potentially attract vital additional revenue, in order to increase the chances of success of the operation.

At the time, statistics from cities across the world demonstrated that there was a real need for more taxis in London. Paris, with a population of 4 ½ million used 16,000 taxis on its streets, Milan, with 3 ½ million people ran 7,000 vehicles, and New York with 8 ½ million people relying on some 12,000 licensed taxis, which figures did not include what was thought to be many other 'unlicensed' taxis, too. London with a population estimated at 10 million, offered a mere 6,000 taxis on its streets. Further questions were already being asked in Parliament on the subject of minicabs. In the Commons on 15 March 1961, Mr Butler the Home Secretary, was asked by Rupert Speir if he knew about the proposal to operate minicabs without a license from the Commissioner of Police, and driven by people, who were also unlicensed. After Mr Speir had explained how their systems would work, he asked if the Home Secretary would introduce legislation to make this method of hiring illegal. Mr Butler said that he had been advised by the Police Commissioner that this system might constitute plying for hire within the 1869 Act. If sufficient evidence was gathered, the Police might consider a prosecution.

In Welbeck Motors' own surviving correspondence, the idea for 'Project Minicab' had been said to have first been mooted in December 1960, a month after Rupert Speir's debate in the House of Commons concerning the question of minicab operations. If correct,

Michael Gotla (Graham Walker – Welbeck Archive)

this demonstrates the swiftness with which the whole project was put together and launched, and Michael was also very clear on his definition of the wording, in a memo of the time he stated 'Please note that the correct spelling is "**Minicab**" one word, no hyphens – when it applies to the Welbeck Motors project'.

With its new financial backing, the other option open to Welbeck at the time would have been to branch out with another Ford dealership, as had previously been discussed by the board; one location had been identified in Streatham, London. Interestingly, Renault had also put forward the suggestion, that Welbeck Motors should undertake the organisation of its distribution for the whole of the London area. Both Michael and Mr Walker were undecided as to which of these two courses would prove most advantageous for the company.

Graham Walker, son of Mr R S Walker, had already spent a considerable amount of time on the Ford project and he was very much on the point of being able to proceed with it. In fact, in a meeting of directors held on 16 May, Michael said that the company had paid a deposit of £150 for a two month option in respect of a showroom, the rent of which would be £2,500 per annum for twenty-one years. The Ford Motor Company had approved it and it was a question of finding suitable workshops in the same area. Sadly, this did not come to fruition, due to not being able to find the much needed workshop space and the option on the lease of the premises eventually lapsed.

However for Michael it was the idea of the new glamorous world of minicabs, perhaps coupled with the appealing thought of the personal publicity which would push him down the path of this alternative project. The choice of suitable vehicles for the venture would obviously be crucial to the initial cost of launching the scheme. When the plans went public, the vehicles initially proposed were either Ford Prefects or Renault Dauphines. However Ford was unable to commit to the project, because it was concentrating its efforts on new models in its range and, aside from this, its two-door model had obvious limitations as a minicab. BMC had also been approached with the suggestion to use Morris Minor 1000 four-door saloons, but with BMC's ties to the taxi trade, through the Austin FX3 and new FX4 models, it followed that it was unwilling to do a deal.

Renault though was much more receptive to the idea, because of the additional publicity, which would be generated from the use of its four-door Dauphine, and, perhaps, remembering that in Edwardian times, Renault had been a major supplier of taxis to most of the worlds' capitals, including London.

Key personnel would also play a big part in the Welbeck Minicab operation's success. The main personalities involved in the project were Mr R S Walker, as Chairman of Welbeck Motors Ltd; Michael Gotla, of course, as Managing Director; then, as joint General Managers, Commander C W B Milner RN and Mr David Evans, the latter being the former taxi driver and Michael's first business partner at the start of his Welbeck Motors days, who had re-joined the business and was to be the number one controller at the minicab base. David, generally known as Dave, possessed 'The Knowledge' and a full understanding of London's streets, which made him the ideal man for the role, and he would play an invaluable part as the operation started. In fact, it was Dave, who was said to have been the key person assisting Michael to formulate the argument to operate legally without 'plying for hire' and he was at Michael's side for the whole time.

Bill Buck and Jane Simpson in a publicity photo
(Graham Walker – Welbeck Archive)

To add to the glamour of the operation, Michael also wanted to recruit female drivers, as he felt some customers would feel more comfortable being driven by a lady. A young 22-year old, former RADA student, Miss Jane Simpson, was to be the public face of the operation and, in pre-launch photo shoots for the press, she modelled the new uniform, along with a male counterpart, Bill Buck, and later, in a driving capacity, as the public face of 'Miss Minicab'.

There were to be day and night shifts manning the phones and radios, which would require at least ten people. It was a genuine twenty-four hours-a-day, seven days-a-week service, with new premises being leased, not far away from the Welbeck Motors car showrooms, at Taunton Place, next door to the 'Boston Arms' public house, just a stone throw away from London's Marylebone Station. Here, a team of radio controllers and dispatchers would operate the switchboards. The offices for the new venture would be based at the same place and the minicabs themselves would be prepared for service, maintained and repaired on these premises when in operation. An on-site canteen would cater for the staff, including the mechanics, cleaners and operational staff. The Crawford Street showrooms were to be kept separate, continuing, as before, with the car sales and rental business. On foot both premises were only a

Welbeck Motors mechanics undertaking the necessary upgrades in the workshop at Taunton Place to the new Renault Dauphines before they enter service in June 1961. (Graham Walker – Welbeck Archive)

The Welbeck Motors workshops are filled to capacity with the new Renaults, which are yet to receive their attention-drawing advertisements. (Graham Walker – Welbeck Archive)

ten-minute half a mile walk distance from each other. In that May, a Mr R G Wilson (who was a personal assistant to David Finnie on the board) was appointed to the board of Welbeck Motors, meaning there were no less than three qualified chartered accountants representing the General Guarantee Corporation's interests.

The minicabs themselves were actually serviced on the floors above ground level at Taunton Place, so the use of an internal lift was necessary to access the workshop area space for this purpose.

A crucial part of the success of the minicab operation was to be the effective use of the two-way radios between minicabs and the Welbeck office. Michael had made contact with Mr Mead of the Radio Services Department of the General Post Office (GPO) on 3 February 1961, when he requested an interview with

A Welbeck Motors mechanic is captured in a brief moment of distraction as work preparing for the vehicles' conversion from 6 to 12 volt electrics progresses. (Graham Walker – Welbeck Archive)

that organisation, in order to discuss licenses and the allocation of channels for the radio telephone apparatus, which he wanted to install in the minicabs. For readers of a younger age, the GPO's purpose as an organisation was a combined function of state postal system and telecommunications carrier, of which the Radio Services Department was part. Mr Mead immediately called the Home Office for advice. Its initial reply was 'While, on present information, the Home Office would not wish to place obstacles in the way of Mr Gotla's application, there might be an advantage in the two departments keeping each other posted about developments'. In view of what he had been told in a discussion, Mr Mead postponed a meeting until Michael had completed 'certain' documents, which, by that time, the Home Office Commissioner's further correspondence might be known.

It was pointed out that, under the procedure by which the GPO licensed private firms and individuals to operate radio telephone equipment, it would not be possible for an intending hirer to make a call himself to the radio cab headquarters, only the licensee and his employees (i.e. the drivers of the vehicles) could use such apparatus. Consideration was also given to whether Welbeck Motors' proposals were lawful and the information was passed to a Mr Pike at New Scotland Yard.

On 16 February, Michael sent the completed application for a Private Mobile Radio system (PMR) to the GPO and advised it that he had consulted Pye Telecommunications Ltd, which would be their suppliers, and, on Pye's advice, it required six pairs of operating frequencies within switchable distance. Michael went on to say that, no doubt, the organisation would have seen the many press notices regarding the proposed method of operation and that he proposed to have a control centre on the premises at Taunton Place, Marylebone, London NW1. From there, two remotely controlled fixed stations would be operated via the GPO landlines, details of which were included on the application form. These stations were located at Crystal Palace, in the south, and Hampstead, in the north. These two beacons were linked to the Welbeck Motors' office by land line. The press reported that this would be Europe's largest private radio station costing £110,000.

The method of operation would be that the driver of the Welbeck Minicab would take verbal instruction from the prospective client and pass this instruction by radio-telephone to the control room, where it would be duly acknowledged, documented and thereafter authority would be given by the control room for the driver to proceed with the carriage of his customer. Michael outlined that the plan was to commence with two hundred vehicles early in June 1961, but this number would increase to eight hundred within twelve months. If the six channels proved to be insufficient to cater for the expected level of radio traffic, Welbeck would require additional channels. The six-channel operation itself was not exactly common at the time where many users would only be allocated one radio frequency.

The GPO had written to 'E' Division at the Home Office on 21 February to inform it of developments and mentioned that it felt that the application was valid, although, in their opinion, the application was optimistic regarding the number of vehicles to be catered for! In essence, it was just another radio-taxi scheme, despite the booking procedure being reversed, and the GPO would allocate frequency channels for it on a similar basis. Unless the Home Office stated anything to the contrary, it would go ahead with the frequency allocation, but the license itself would not be issued until the system was ready to go on the air. By now, the involvement of the Home office had reached a particularly high level. Sir Henry Austin Strutt KCVO, CB, MA, JP, (1903-1979), who worked as a senior civil servant in the Home Office between 1925 and 1961, and who was Deputy Under-Secretary of State between 1957 and 1961, had expressed the view on 10 March that it would be most undesirable for the GPO to grant facilities for minicab operations, which were contrary to the law, and he suggested a meeting with the GPO about the situation. The GPO wrote to both Welbeck Motors and to Carline on 17 March referring to their applications for radio telephone facilities, and it drew attention to the Home

Secretary's statement in the House of Commons on 15 March, in which he warned that this proposed method of operation might be illegal. It did however state that it would reserve two channels in the VHF band for the initial development of the system, but it pointed out that a private mobile radio license would not permit passengers or prospective passengers to make calls over the radios in the vehicles, which apparatus could only be used by firm's employees and the licensee. In their reply, Carline explained that it was their intention to incorporate the radio link with their normal pre-booking car service and that the radio would not be used by any vehicles, which might be construed as plying for hire. The apparatus would only be used by their staff and not by members of the general public. Michael Gotla observed that the company was conscious that private hire cars were not allowed to ply for hire, yet that the law appeared to be very confused as to what 'plying for hire' actually was. His companys' intention was to revert to normal private hire procedure, if their chosen method of operation was shown to be illegal, and that their intention was to operate strictly within the law. But, as the law on this particular point was obscure, Michael suggested that a test case might be necessary for clarification.

On 24 March, the GPO asked the Home Office for its comments concerning the two firms' replies, because, on the face of it, it felt that Carline's assurances could be taken as such, but felt that an early test case would be desirable, in order to determine the legality or otherwise of Welbeck Motors' venture. It felt that there was no point in holding up the application any longer. Internally, the Home Office suggested that it would be acceptable to raise no objection to the issuing of the necessary license; however, its legal adviser had expressed a view that the proposed method of operation was contrary to the law, but that it was not possible to obtain an authoritative ruling without bringing the matter before a Court of Law. It was felt that if radio facilities were refused to Welbeck Motors, there would never be an opportunity for the Home Office to gather evidence, which might be used to test the legality of the operation in future, nor would Welbeck be able to operate as efficiently as it wanted, as a private hire business. With this, a suggestion was made that no objection should be raised. On the 4 April, Mr Mead from the GPO wrote and agreed that it did not wish to refuse a license to either firm, but, in the case of Welbeck Motors, he merely pointed out the legal position and that the Home Office were aware of the background to the application.

As a footnote concerning the GPO, it was founded in the reign of King Charles I in 1660, and it was finally abolished in 1969, whereupon its assets were transferred to the Post Office, when it became a statutory corporation.

Michael was in the process of obtaining final estimates for the various costs, in which the setting up of the radio service would be involved. One surviving letter, although not on headed paper, but probably from Pye, the chosen suppliers for the operation, was sent to Welbeck Motors on 15 March 1961 and it provides an insight into the many diverse costs of setting up the operation.

For example, the costing includes rental maintenance terms over a five year period, in which were included free replacement of all faulty components, labour costs, replacement valves, regular overhauls and a prompt repair service for the equipment. The base station equipment comprised eight PTC 753DV fifty-watt fixed transmitter/receivers for two frequency Simplex operation, complete with main/stand-by changeover panels @ £10.16.0 each per month.

Also, there were the four PTC 411/458 remote control units and panels with main/stand-by switching (for connection to GPO landlines) at a cost of £2.14.6 each per month and the unit cost per week per mobile for fixed station equipment was 2/3d. Information from the Pye History trust suggest these were 50 Watt PTC354 fixed station transmitter-receivers.

As for the mobile equipment, a quantity of 200 PTC 6/2007V/12 Ranger mobile sets with six channel switching facilities would be required and they would be supplied with crystals for one-channel operation to operate from a twelve volt DC vehicle battery and having two-frequency Simplex operation. All installation parts would be charged @ £3.15.0 each per month and with a unit cost per mobile per week of 17/4d. The communication radius between the minicab and the base was said to be in the region of twenty to twenty-five miles.

Engineers at Pye preparing and testing the soon-to-be-installed radio sets for fitment inside Welbeck Motors' Renault Dauphines. (Graham Walker – Welbeck Archive)

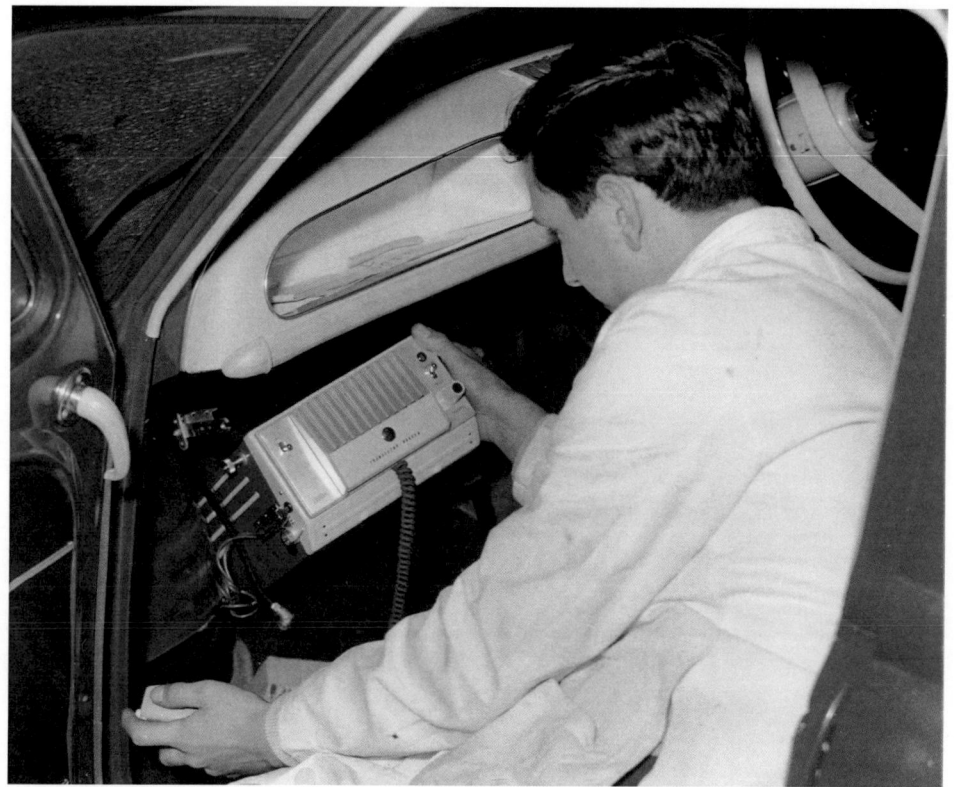

An engineer from Pye is seen installing a new two-way radio set in a minicab ahead of the launch day. (Graham Walker – Welbeck Archive)

The Pye dash-mount PTC 2000 Ranger family of Private Mobile Radio (PMR) mobiles was the then main medium power mobile product line in production at Pye from 1956 until about 1963 and was supplied in the tens of thousands to Business and Industry users world-wide in a number of versions, with both under-dash mounting and remote boot mounting available.

The PTC2007 so-called Transistor Ranger was a later model using transistors in the power supply to generate the high voltages needed for the valve technology of the time.

The installation costs of the fixed station equipment, of the remote control equipment and the supply of fixed station aerials, feeder cables and commissioning would be £350.0.0.

Installation of the Ranger mobile sets in the Renault Dauphines, based on two hundred cars, would be £600.0.0.

For rental of the facilities at remote sites in Crystal Palace and Hampstead for the operation of two fifty watt fixed stations and two stand-by stations on each site, including electricity, would be £208.0.0.per annum.

The GPO licence fees, payable to the Radio Services Department of the GPO, for four fixed transmitter/receivers and two hundred mobile sets would be £410.0.0 per annum.

Inside the Taunton Place workshops, an engineer is about to install a Halda Taximeter inside a minicab. (Graham Walker – Welbeck Archive)

Installed Pye radio set and Halda taximeter in their interior
positions (Graham Walker – Welbeck Archive)

And, lastly, the GPO landline rental, payable to the Telephone
Manager, Centre Area, for the rental of four GPO private wires, two
from Taunton Place, and two to Crystal Palace, with a total direct
distance of twenty-three miles, would be approximately £235.0.0
per annum.

In practice, two channels were available at the start of operations:
the red channel was for the operational base and the blue channel
for the cars, with the vehicles each being numbered.

In early stages of the system being used, a lot of audio interference
was encountered at the Star Sound recording studios from the
minicab operation, which was located from a frequency point of

view, less than one mile away from the Taunton Place base and required further filtering at the base stations to stop the interference when the minicab operation would be fully functioning.

It must be noted that in the studios long history, first opening in 1937, it had recorded for performers such as Arthur Bass, Max Miller, Jerry Hoey and his Band and after becoming Audio International Studios in its forthcoming years would be affiliated with producer George Martin and recording artists such as The Hollies, Mud, Gerry Rafferty, Suzi Quattro, Pink Floyd, Gilbert O Sullivan and Leo Sayer.

Chapter 3

Publicity & the Launch Day Arrives

At this stage of the build-up to the launch of the minicab operation, Michael was, indeed, in his element. Ever the showman, he was in great demand with the gentlemen of the press and, one suspects, he was enjoying what would be the peak of his time in the limelight.

As word continued to spread of the new plans for a minicab operation, Michael became busy with articles in and interviews for the press, which were fast raising the profile of what would happen in the June of 1961. It has to be said that there could have been hardly a newspaper across the country, which was not running features of some description or another, together with details on some aspect of the operation. Reports suggested that something in the region of £480,000 had been invested into 'project Minicab', an operation on a scale, which had not been witnessed hitherto. Ahead of the launch, other press reports suggested that a director of the corporation, Mr David Finney, had stated 'that the General Guarantee Corporation's total direct financial investment in the company at that point, including loans, was well under £250,000'.

Over the following weeks, many stories in the press kept readers informed about how the position of the traditional London taxi driver was seen to be under serious threat from this new competition. In the 'Evening Standard' newspaper of Friday, 9 December 1960, the headlines of Robert Walling's article entitled 'Car Talk', read, 'As easy as taking a taxi – the new way to hire a car' and asked 'mini-taxis for London? Not yet. Small car taxis? Yes, if Michael Gotla has his way'. He was described as 'a fast rising, West End car hire and sales' man. The article went on to describe his plans for minicabs, and to reveal something on his background. Interestingly, the article mentioned that Michael had taken their reporters up to Lord's Cricket Ground, where hundreds of stored Welbeck Motors' cars and vans were parked at the Nursery End. The interviewer said that 'for this faintly blasphemous facility, he pays £800 per month in the winter' at a time when the motor trade is worried about car

WOLFSON BACKS MINI-CABS
Mr. Isaac Wolfson, head of the Great Universal Stores Group, has put £1,000,000 into Welbeck Motors, the West End mini-cab firm which is to put its fleet of 200 cars on the road on Monday week. (Page Two).

Wolfson backs minicabs. (Graham Walker – Welbeck Archive)

TERENCE DONOVAN

Michael Gotla. *Began as ad agency copywriter by day and driver of hired car by night. Opened own car-hire business with two (hired) cars and slowly expanded, hiring, leasing and selling cars, until Isaac Wolfson backed Welbeck Motors and his Mini-cab project, and Gotla's fleet of two grew into something nearer 4,000.*

Michael Gotla, 42 (St Paul's and UC), who built up Welbeck Motors, insists 'I'm basically a lazy person without any feeling of ambition.' But with an ear-splitting laugh he mentions the kicks he gets out of business – like having his own army of chauffeurs causing that fantastic publicity and enormous excitement in France, getting his fleet of Renaults for an astonishingly low price and giving Ford's a bitter pill to swallow.

Determined, he went ahead with the Minicab project, despite powerful opposition – both clumsy and bitter – from the London taxi trade. 'It's simply never any good taking "no" for an answer. For example, if you can't get an overdraft and the bank manager gets shirty, you ask to see his superior and you get results. So many people lack this quality or don't realise it's possible. 'The Minicab project excites me – it brings out the essential backroom boy in me! I prefer creative advertising to power; and the selling policy of this business has always been advertising!'

His main regret about the rapid growth of the business in recent years is that he's no longer 'one of the boys, doing the town together, having fun. A bachelor can find an astonishing depth of friendship with his staff because he is with them so much. But now, I've become a slave to the business – aged ten years in the last few months.' One of his favourite pastimes is pushing off to a local church for a couple of hours when he can spare the time, 'just to sit there when there's no service. I don't like the clergy, but the atmosphere is peaceful and the architecture beautiful.

'Money doesn't really interest me. From the age of nineteen I thought the world owes me a living. I've always lived comfortably – that's why I did two jobs at the beginning, copywriting by day, driving at night. Since I started Welbeck I've had my own nice flatlet upstairs, and a house at Rustington by the sea for week-ends. A year or two ago I converted an old Georgian house in St Albans into a show-place, but it's too posh for me now. No, I'm not a rich man and my tastes and possessions haven't really changed. Expense accounts? I just don't believe in lavish business entertaining. When I want something, I go and see the person.

'When I've got sufficient money, I think I'll get out of business and buy myself a big house in the country, make it into an orphanage and surround myself with kids – let 'em enjoy the things I like, the sea and boats.'

sales. However, an unworried Mr Gotla said 'Sales are picking up again. We sold fifty-seven new Fords in the last forty-two working days'.

On 31 January 1961, the 'Financial Times' asked 'An increase in taxis?', in which its reporter took a more in-depth view of the costs of operating traditionally licensed taxis in the capital, and the probable financial impact of Welbeck Motors' venture, with its much reduced pricing structure. The reporter was sympathetic to the plight of the taxi drivers, citing their high operating costs as a significant factor for the larger taxi firms. The article indicated that, initially, the industry's new Austin FX4 model had not been as well received by the taxi trade as had its predecessor, the FX3, and that Beardmore Motors, which, at the time, supplied about 10% of all London taxis, had secured an arrangement to introduce a new model that year. Even so, it was felt that the big news of 1961 would still be made by Welbeck Motors.

On 21 February, the first of what would become many television interviews, took place on BBC TV's 'Tonight' programme, which ran an interview made earlier that day with taxi drivers and with Mr Francis of the Men's Union, on which programme their views were voiced. Michael was also interviewed, and the interviewer, Macdonald 'Mac' Hastings, asked how Michael would 'get around' the so called 'Snag' (i.e. that of minicabs being hailed), as there were regulations controlling taxis, and, under those regulations, a minicab could not be hailed. Michael replied, 'I don't like this "get round it". We have found a perfectly legal way of doing it. Our cabs are going down the road and you can hail one; it stops and you give your name to the driver and tell him where you are going. He relays that information to our control at headquarters, where they note

Mr Isaac Wolfson 'Millionaire backing Welbeck Motors' is seen arriving at the Wolfson Institute on 3rd May 1961. (Keystone Press – Author's collection)

it down. We consider that such an action constitutes an advance booking; the driver then says 'get into the cab, sir', and off you go".

Over the next few months, mentions about and actual interviews of the scheme were to be aired on the national BBC News, on ITV, on 'Town & Around', on 'Any Questions', on 'In the South East', on 'This Week', on Radio Newsreel, on 'Today', on 'News Extra' and on 'Changing Face of London' programmes, such was the media interest.

The 'Steering Wheel', the twice monthly journal of the British taxi industry, had been keeping its readers informed about minicabs firms' developments and, in particular, of Welbeck Motors' plans. After much thought, Michael had written a letter to that journal which was published in the 27 May issue and which gave his views, in the hope of correcting some of the misapprehensions and to relieve tensions. His lengthy, detailed and diplomatic contribution was followed by a letter from Mr John ET Welland and Mr J H Francis of the London Motor Cab Trade Joint Committee, who offered their opposing views adjacent to Michael's letter.

On 7 June 1961, the 'Western Daily Press & Bristol Mirror' published 'The Man behind the minicabs is named'. The reporter said that there is an end to speculation about the source of the finances of Welbeck Motors. Mr Isaac Wolfson, of the private company General Guarantee Corporation, was revealed as the 'man with half the shares'. Mr David Finney, the executive director of General Guarantee Corporation said that, although they had control, they did not run the day-to-day business. He was further quoted as stating that their direct investment in shares and loans was less than £250,000. 'Renault Magazine', the French company's house journal, in its June/July 1961 issue provided a two-page illustrated

A portrait of Michael Gotla by Rex Coleman. (National Portrait Gallery)

spread of the first day of operation in French text, with pictures showing a smiling female driver leaning out of her driving seat and another of a male driver with his hand on his car's two-way radio, ready to accept a fare. A Dauphine was also captured threading its way through London's busy streets with London buses and black cabs in close proximity.

Michael was certainly not shy of publicity, as the media interest was gathering pace ahead of the July launch of the minicab operation. On 9 May he found the time to sit to have his photographic portrait taken by the noted photographer Rex Coleman (1937–2006), of Baron

Studios, who had captured the likeness of many of the celebrities of the day. Mr Coleman was appointed official photographer for the National Photographic Record at the National Portrait Gallery in 1967. In 1999 this image of Michael was donated by Godfrey Argent to the National Portrait Gallery in London, where it is to be found to this very day.

On the 1 June, a conference at the Welbeck Motors' showrooms was held for the benefit of the 'London Daily, Evening and Technical Press' in order to outline the intentions of the operation. It was confirmed that, on 19 June, the initial fleet would be two hundred strong, and this number would be increased rapidly to five hundred units. If the pilot scheme were to be successful, ultimately two thousand vehicles were envisaged. Each vehicle was required to earn £6 per day to break even. The drivers would retain one third of the meter reading, with a guaranteed minimum of £7 per week. It was stated that the Norwich Union Insurance Co. had extended the same cover to minicabs as it did to licensed taxis. Michael also stated to the press that it was quite legal to use a minicab as a mobile booking office, but 'as a peaceful gesture to taxi drivers, anyone hailing a minicab would be asked to wait for a second cab to pick them up'.

Discussion followed about the radio system, about the fact that British Railways would welcome minicabs at its stations, and about the advertising scheme, with Michael concluding that he hoped that his minicabs would come to be accepted quickly as part of London's overall transport system. But, he said, he could not afford to lose money on too large a scale with his experiment and London, with its terrible traffic jams, might send the whole undertaking haywire. After three months, he felt he would know more clearly whether to go ahead in a very big way in London, or to concentrate on the provinces. After this, there was an opportunity to be taken out in one of the as-yet-not-in-service Renault Dauphines, and to participate in a conducted tour of the radio control room and servicing department at the Taunton Place premises.

In due course, there would be the further facility for petrol pumps to be installed at the building when it was completed; it was expected that a full tank of petrol in a Dauphine would easily last one full shift in service. Once operations were under way, the National Benzole Company Ltd offered Welbeck Motors a loan of £18,600 over a twenty year period at 5% interest, in order to finance the installation of the new fuel tanks and pumps at Taunton Place, which offer was accepted by the board, as this installation would greatly assist in the monitoring of fuel usage.

In early June comments were made at Westminster, when a dinner party 'arranged by Mr Ronald Bell, Conservative MP for South Bucks' was held at the House of Commons. He wanted MPs to be able to meet both Michael, himself and his PR firm Michael Clark-Hall, of which Mr Bell was an advisory director. In a statement to 'The Sunday Telegraph', Mr Bell commented that, with the new service being so controversial and such a live political issue, the directors wanted to meet MPs, who were interested in the subject. Mr Bell had been asked if he could help Michael and his directors to meet politicians from both parties and several MPs and peers were interested. At dinner, an amicable discussion took place, during which many controversial questions were put to Michael and his directors, and this meeting was believed to have helped both sides to appreciate the problems involved.

On 14 June, Michael's solicitor from Amery-Parkes & Co wrote to him, ahead of the minicab launch, in order to summarise his concerns about the current situation, which had arisen from their discussions. Firstly, the proposal of a hirer telephoning the base and a minicab being routed to them could not be seen to be 'plying for hire', although he felt that 'it could be said to be a little nearer to plying for hire than would be the case if the cabs were to be sent from a garage, as distinct from being already on the streets'. The second proposal where a hirer could stop a minicab, the driver of which would then telephone headquarters and get confirmation of the booking, would, in his view, certainly be held to be 'plying for hire', and that a prosecution against them could not fail. The third method that had been suggested was where a hirer stops a minicab, the driver of which then telephones for another minicab to pick up the passenger. In this case, in the solicitors view, there was some chance of it not being 'plying for hire'. This opinion was based on a case at the Divisional Court, Cogley v Sherwood in 1959 about London Airport, where the Lord Chief Justice had stated 'For myself, I think it is the essence of 'plying for hire' that the carriage should be exhibited'. With this in mind, Michael's solicitor felt they had a strong arguable case.

Another point, into which he had been looking was Michael's idea of forming a club for the venture and he was in the process of preparing some draft rules. He felt that it could be successful, although, previously, tour operators had run into some difficulties, in a case brought before the courts against Millbank Tours. They had been arranging continental air tours for members of a particular association or club, but not for members of the public, which was forbidden by the Air Corporation Act 1949. Nothing in the Act

Crawford Street, London. It's early June 1961 and a small selection
of the new Renault Dauphines display 'on test' stickers in their side
windows as the launch day will soon be upon them. A curious member
of the public looks on. (Graham Walker – Welbeck Archive)

related to hackney carriages, and the fact that they were plying for hire meant that they would be available to the public.

Finally, there were two other outstanding matters, on which Michael had asked for advice. One was on the subject of advertisements (more about this in Chapter 5) and the other was on the Royal Parks (see Chapter 8). Michael was well aware that, in the future, prosecutions could be expected, and, with this in mind, his solicitor advised him to obtain representation from Mr Gerald Gardiner QC, and a junior; the fees alone for two hearings could well be in the region of £1,600, which, in his opinion, required some thought.

Three days before the start of operations on 16 June, the use of minicabs themselves was in the press, but for quite another reason.

An interested member of the public inspects this new form
of transport, parked in Crawford Street, to see what it could
potentially offer. (Graham Walker – Welbeck Archive)

This time it was because a film was being made for French Television
and the director of the shoot thought it would be a good idea if nine
of the minicabs were to be lined up outside the House of Commons.
The drivers duly did what was instructed of them and they ran
into trouble with Scotland Yard's traffic police department, when
they were observed travelling in line, at a much slower speed than
the other traffic, whilst being filmed heading towards Parliament
Square. The problem was that the advertisement emblazoned cars

On a rainy June day, a Welbeck Minicab driver familiarises himself with the two way radio whilst being photographed for publicity purposes. (Graham Walker – Welbeck Archive)

A more close-up photograph, clearly showing off the minicab
advertisements. (Graham Walker – Welbeck Archive)

were being parked within a three miles radius of Charing Cross
station and as the archaic law stood, at that time vehicles could
not be used for purposes of advertising within the stated area. The
minicabs were being both parked and used for publicity purposes
and in this instance, it was deemed that they were breaking the
London Waiting and Loading Restrictions Act 1958. The drivers
were questioned by the police and they were advised that a report
would be made to the Traffic Department.

Eventually they were summoned to Bow Street Court on 27
September and, after pleading guilty, they were all given absolute
discharges. Welbeck Motors were fined 50s. on each of the nine
summonses and ordered to pay seven guineas costs.

The day prior to the launch, the drivers at their 'school for minicab drivers' were using their time to have a dress rehearsal, to try out the radio system and to familiarise themselves with what they could expect when in service. Drivers were told 'Always refer a customer, who hails you in the street and who wants a cab for a short distance, say from Piccadilly to Knightsbridge, to a cabbie. That is the job for him and not for us. There is no taxi war as far as we are concerned'. 'Miss Minicab', Jane Simpson had had a private dress rehearsal during the week. She said that, in Belgrave Square and Piccadilly, she found that taxi drivers were inclined to cut her up, but, when they saw she was not aiming to do battle, they got on with their job and let her get on with hers.

All minicab drivers were to receive a uniform, which gave the press the opportunity to name the staff 'Gotla's Army' or 'Fidel Castro', as it was based on American lines, with shirt and trousers in a Bedford cord woven cotton, which was drip dry, minimum iron, spot proof, colour fast and water repellent. They were olive green in colour and manufactured by Gardiner of Aldgate, and supplied through a Mr Peacock of the Great Universal Stores. To finish the look, these came with a soft peaked hat in similar material and a black tie.

At the end of May, on the steps of the Eros Statue in Piccadilly Circus, the uniform had been officially launched ahead of the start of operations, when it was modelled by new drivers Bill Buck and Jane Simpson, who posed alongside a minicab. Apparently, while the photo-shoot was underway, a passing traffic warden placed a parking ticket on the minicab! Images of the models were circulated by Welbeck Motors' PR Company and published in many newspapers across the country, including the 'Manchester Evening

Advertisement placed in the Evening Standard Newspaper ahead of the launch day of the minicab operation. (Graham Walker – Welbeck Archive)

42 — EVENING STANDARD, MONDAY, JUNE 19, 1961   ADVERTISERS' ANNOUNCEMENTS

50 years on . . .

Left, the Renault taxi of 1907. Above, part of the new Renault Minicab fleet.

These little cars
—ideal for the job

RENAULT cars have now gone full cycle as part of London's transport. More than half a century ago in 1907 Renault's built the first taxi cabs ever to ply for hire. (See picture.)

Today London's latest fleet of hire cars, the Welbeck Minicabs, which make their debut this afternoon (Monday), have all been built by Renault.

Two hundred will be on the streets today. In the next few months there will be 500.

AT THE MARNE

RENAULT'S will be adding another page to their long history. For the victory of the Lattle of the Marne was made largely possible by the help of Renault taxis in transporting troops and supplies.

For the past few weeks regular travellers on the London-Southampton road have become used to the

Inside the new Minicabs . . . a comfortable ride assured.

sight of great articulated car transporters full of little red Dauphines bearing the legend:

"More Minicabs for Welbeck Motors."

These little cars have already had an exciting journey.

Produced at Flins, the ultra-modern factory 40 miles from Paris where all Dauphines are built, they have travelled on special car transporter trains to Le Havre.

There they have been loaded on to the Liberty ships of the Renault fleet to cross the Channel to England.

At Southampton they have been off-loaded into Renault's

LONDON'S NEW TRANSPORT

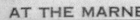

PAGE TWO

depot right by the quayside.

Up till now they had looked a trifle dull, for their gay paint and bright chrome were covered with a protective film of wax to keep them safe against the sea air and spray on the Channel.

This coating is removed at Southampton and like butterflies from a chrysalis they emerge — "Red Admirals" to flit around London.

IDENTICAL

THEN they started on their final journey to London.

And none had covered a mile on their own wheels until they were delivered to Welbeck Motors.

Much midnight oil has been burned in Flins to get this fleet delivered on time. For even to a factory geared for mass production, such a fleet order to be delivered on time, every car identical in colour and inside trim, puts on the pressure. But they are here. And from this afternoon they are at the service of the public.

The Dauphine is specially

suitable for the new job it has been chosen to do.

To begin with it has four doors. The rear-mounted engine brings the seat forward in relation to the wheels. So the rear-wheel arches do not limit the space of the back seat or the size of the rear doors.

Because the engine is at the rear there is no propeller shaft tunnel dividing the floor space. This makes it easy for both rear seat passengers to enter by the nearside door—a great safety factor when getting into the Minicab in crowded traffic.

COMFORTABLE

A COMFORTABLE ride for the back-seat passenger is assured by the car's special suspension—a combination of coil springs and sealed air cushions which give the same springing whether the car is laden or unladen.

The Dauphines used as Minicabs have a four-speed gearbox and an 845 c.c. motor-cooled engine and independent suspension on all four wheels.

The boot under the bonnet has a capacity of seven cubic feet and can be locked and unlocked from the driver's seat.

Overall length of the Dauphine Minicab is 12ft. 11in. and its width 5ft., so it is a compact car for town driving.

Its turning circle approaches that of the conventional London taxi—just under 36ft.

A Renault Dauphine does about 40 miles to the gallon in heavy traffic conditions and 50 on a long straight run.

Peter Jackson

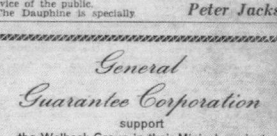

ADVERTISERS' ANNOUNCEMENTS   EVENING STANDARD, MONDAY, JUNE 19, 1961—13

What they think of Minicabs

PICTURE PROBE asks the public

MR. WYNDHAM WILLIAMS, a stockbroker who works in the City and lives at Horstead Keynes.

"It's just a matter of simple economics as far as I am concerned. I have to take cabs, I haven't a limitless income (who has?), therefore when I have to get from one part of London to the other I will, from now, have two 'markets' to shop in."

LONDON'S NEW TRANSPORT

PAGE THREE

MISS LESLEY McCLURE, an Australian who lives at Muswell Hill. "Back home in Melbourne our cabs are much smaller than the 'standard' ones you have in London. I am looking forward to my first Minicab ride very much. I am not a 'taxi-addict' but when I do take a cab I expect value for money — and civility.

"I have found the ordinary cab driver very helpful indeed when I have been in a jam. Of course minicabs being cheaper than the others is going to be an important factor for a girl like me, when I have to get somewhere in a hurry."

PATRICK JOSEPH LISTER, a newspaper seller who has been standing outside St. Paul's for so long that, as he says, "I've become a blooming landmark."

"We are going to see a lot of changes and I'm not quite sure if they're going to be far the better. I know hundreds of cabbies, both owner-drivers and journeymen drivers, and I know what stiff tests they have to go through before they're even allowed on the road. I can see trouble coming up but maybe it will sort itself out. I think the public will still be faithful to the old-time cabbie. American visitors love 'em."

MISS BABS MERCER, a secretary in the City who lives in Holland Park. "I'm all for Minicabs. I think they are a jolly good idea. What my dog Adam is going to think of them, though, is another matter. He adores riding in cabs—he always has. They tell me there will be some women drivers. I'm looking forward to that if it's only because of the novelty of it all. And anyway all of us can do with more colour in the streets."

Last word from a cabbie . . .

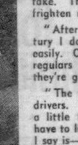

Taxi-driver ALBERT PEAGRAM, who has been "pushing a cab round London for over 25 years."

"Well, they're here and we will have to learn to give and take. They aren't going to frighten me one little bit.

"After a quarter of a century I don't get scared very easily. Of course I've got my regulars and I don't think they're going to desert me.

"The same goes for other drivers. There's bound to be a little trouble but we will have to live and let live. What I say is—let 'em all come."

ADVERTISERS' ANNOUNCEMENTS

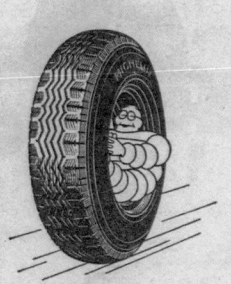

For comfort and safety
MICHELIN TYRES
are
standard equipment
on
Renault Dauphines

★

Michelin Tyre Co. Ltd. Factories: Stoke-on-Trent (Head Office) & Burnley

B61/33/M

Ready . . . for a call every two seconds

A control room which the makers, Pye Telecommunications, Ltd., claim will be Europe's largest radio-telephone system, will be the nerve centre for the Minicabs. It has cost £110,000 and will go into action this afternoon.

From this control room the 200 Minicabs will be in constant radio communication in an area covering 25 miles of central London.

Two new transmitting stations have been established by Pye, one at Hampstead and the other at Forest Hill.

For the past eight weeks radio experts have been busy installing radio-sets in the 200 Dauphines which take to the road this afternoon.

These sets are the latest Pye equipment and are partly transistorised.

50-mile range

IN charge of the new installation at Welbeck's Taunton Place garage is Pye's engineer, Douglas Keating.

He said today: "The system has been planned with six-channel radio - telephone equipment. At first only two channels will be used but when the fleet is increased to 500, four will come into operation.

"With good line of sight conditions the actual range of the transmitters will be nearer 50 than 25 miles."

In the special control room, 20 GPO telephone lines have been installed and a "conveyor-belt" message system

LONDON'S NEW TRANSPORT
PAGE FOUR

has been devised to handle calls at the rate of one every two seconds.

Converted

PYE'S are specialists in supplying taxi radio schemes and their installations include fleets of taxis in Montreal, Paris, Mexico, Milan, Rome, Beirut, Bermuda, and Brussels, as well as in Australia and New Zealand.

Normally the Renault Dauphine has a six-volt electrical system.

This has had to be converted on the Minicabs to 12 volts, to accommodate the radio equipment. This work has been done by Lucas, the car electrical firm.

Checking the radios which will keep the Minicabs in touch with the control room. Constant communication will be kept with cabs in an area covering 25 miles of Central London.

BEHIND THE WHEEL . . .

THREE HUNDRED drivers have so far been engaged for Minicabs by Welbeck Motors. They have come from all sorts of jobs, but they have one thing in common. They are all experienced drivers.

Here are some of them you might meet when you hire a Minicab.

Mr. Oliver Cromwell Davies, aged 53, married with two children who lives in Stratford, E.15. He has been a long-distance coach driver for 25 years.

Mr. John Woodings, aged 28, from Kingsbury, N.W.9, left the Royal Navy in 1958 and since then has worked as an electrician driver, and is a member of the Institute of Advanced Motorists.

Mrs. Joan Carpentier, a widow of 38, who lives at Queen's Park Court, drove an LCC ambulance from 1943 to 1946 and was a United Transport driver from 1947 to 1950.

From 1950 to 1956 she drove her husband who had his own business.

Mr. Thomas Weeks, aged 45, married, from Courtfield Gardens, Earl's Court, S.W.5, for 11 years worked as a London and Continental tour guide.

Mr. Frederick Davies, of Roseland Place, Ladbroke Grove, W.11, married, aged 25, has worked as a London van driver for five years.

Mr. Barrie Phillips, of Westbourne Crescent, Paddington, aged 27, married with two children, was a driver respresentative for seven years.

Mr. John Smith, of Denning Road, Hampstead, aged 28, single, has driven for a firm of printers and wholesale newsagents for six years.

Mr. Lawrence Ullah, from Mile End, aged 22, married, was two years in the Merchant Navy, and has spent three years as a London van driver.

The Launching of the Welbeck Minicab Fleet

19th June, 1961

You have been asked to come to the Welbeck Building. As soon as you have arrived, you will find yourself in a Welbeck Minicab which will take you to the party. It is only right that you should be one of the first people to travel in a Minicab . . . .

When you arrive at the Minicab Building, you will be welcomed by the Chairman, Mr. R. S. Walker and Welbeck staff will show you the layout of the launching party.

Every Welbeck-man will carry a blue cornflower in his button-hole: so you will know who to ask the way.

We hope the weather is fine, because we want you to find your way to the roof of the building: the focal point of the party.

Timing and programme is as follows :—

11.15 a.m.
Guests arrive. Coffee or drinks will be served on the roof garden.

12.00 to 1.30 p.m.
Buffet lunch on top floor of building. Continuous service during this period.

1.30 to 1.40 p.m. approx.
Fanfare of trumpets calling guests to the opening ceremony on the roof garden.

1.45 to 2.15 p.m.
Opening ceremony ending with the lanching of the fleet by *M. Guitton, Chef du Cabinet, Ministere de l'Industrie et du Commerce.*

2.15 p.m.
Parade of the Minicabs.

2.15 p.m.
All Welbeck Staff (except Minicab staff) start their party.

3.00 p.m.
Building must be clear by this time, because Welbeck Minicab Fleet is in operation.

We hope that you will look round the radio room. We hope that when the opening ceremony has been finished, you will watch the parade of the Minicabs from the roof. We are sorry that the building is half finished, but we thought that it would be so much more fun to have the party in our building than at a hotel. Down one side of the building, we have removed the scaffolding to give you an idea of what the whole place will be like when the re-building is finished.

For 400 guests the lavatory accommodation is sadly inadequate, particularly because this particular part of the re-building is unfinished. But we have done our best: notices will direct you to " Ladies' " and " Gentlemen's " rooms, and the public house next door have kindly given us use of their facilities for the day.

This is the formal launching of the Welbeck Minicab Fleet. But more important than that, it's a party. Many of our guests are from France: the French think that English parties are deadly dull. This time let's prove them wrong.

Programme of the launch day for invited guests.
(Graham Walker – Welbeck Archive)

Some of the Welbeck Motors senior team, ahead of invited guests, arrival on 19 June 1961. Mr R S Walker is second from the right, Michael Gotla, far right. (Graham Walker – Welbeck Archive)

News' and 'The Birmingham Post'. In some cases, the style of the uniform was not helped by the fact that there were only standard sizes available, the cut of these not being suitable always for some of the female drivers. In reality, the uniform only lasted for just over the first year of operation and in the early days of operation, it was specified that the uniform must be worn whenever a driver was driving. Furthermore, it was also a requirement that drivers should not smoke when driving with a customer in the City, although on a longer run, they were permitted to do so, if the customer had been asked first.

On Monday 19 June 1961, the first day of the minicabs operation, Pathé newsreel cameras for television and cinema were on hand both at Crawford Street and at Taunton Place, where the drivers were to be filmed, assembling with their new Renault Dauphines in order to depart to collect their first fares. Scenes from surviving footage show the new drivers walking towards the cameras *en masse* wearing the new staff uniform and all smiling, before getting into their new minicabs to depart for business. In another scene, the booking clerks on the switchboards inside the control room at Taunton Place are seen to be busy taking the telephone enquiries from the public wanting to book these new minicabs; the orders are

This excellent shot shows off the Taunton Place premises, which has yet to see the completion of the working fuel pumps on its forecourt. The roof top party is underway. (Graham Walker – Welbeck Archive)

The Boston Arms public house next door is no doubt enjoying brisk trade on 19 June 1961. Note the policemen present amongst the film crew. (Graham Walker – Welbeck Archive)

then seen being passed to the dispatchers on the other side of the conveyer belt system to be radioed to the drivers. One scene even showed Dave Evans busy on a call, whilst other footage shows the minicabs in service, driving down the streets to collect fares and displaying an obvious presence on a busy local scene.

With his normal flair for entertaining, Michael was at the Crawford Street premises early on the launch day, where he was in great demand for interviews. His public relations (PR) man had given warning that any interviews must be short, fearing that Michael might be near to nervous exhaustion, bearing in mind that he was said to have been up all the previous night, supervising

Taunton Place is filled with brand new minicabs soon to take to
the streets of London (Graham Walker – Welbeck Archive)

Welbeck Minicab drivers wait patiently for the scheduled
3pm start of operation, as the cameras roll to record the
day's events. (Graham Walker – Welbeck Archive)

Michael Gotla draws a crowd of onlookers whilst being interviewed
by the media. (Graham Walker – Welbeck Archive)

last minute arrangements. At the end of an hour, the PR man retired
from the scene, mopping his brow. Yet, half an hour later, Michael
was still talking (some say at a steady rate of three hundred words
a minute) explaining about himself, about the firm and about his
plans for future development. In fact, he was just warming up, such
was his energy. Just after this, a roof-top launch party at the Taunton
Place premises was held by invitation only. The four hundred guests
were diverse, but included the managing directors of Pye and of
Halda, members of the British Motoring Press across the country,
the Continental and American press, the international television

71

On the launch day, a Welbeck Motors' Ford Thames car transporter has arrived near to the Taunton Place workshops, ready to unload yet another delivery of new Renaults. (Graham Walker – Welbeck Archive)

news services of NBC, of Newsweek, of the BBC, of ITV, of Radio Luxembourg, as well as representatives of the national French, Columbia and Canadian Broadcasting services, of Radio Télévision Française, of 'Life Magazine' and of British Newsreels, to name but a few. They were requested to arrive at the Welbeck Motors building in Crawford Street for 11.15am, from which location they would be taken by a new minicab to the other premises. On arrival, they were greeted by Mr Walker and Welbeck Motors staff (all identifiable by wearing a blue cornflower in their buttonhole) and guests were shown around the building, including the radio room and the area where the minicabs were being prepared for service. Between 12.00 and 1.30pm, a buffet lunch was served on the top floor of the building.

Outside the Crawford Street Car Sales Showroom, activity is
observed by a film crew, as minicab drivers stand and watch
the busy local scene. (Graham Walker – Welbeck Archive)

Invited guests mingle ahead of the official launch of minicab operations on
the roof top at the Taunton Place headquarters. Michael Gotla can be seen to
the right from the centre of the image. (Graham Walker – Welbeck Archive)

The Band of the Kings Own Royal Border Regiment provides a
fanfare for the event. (Graham Walker – Welbeck Archive)

Following this, a fanfare of trumpets, provided by the band of
the King's Own Royal Border Regiment, ushered the guests to the
opening ceremony on the roof garden, to which a Renault Dauphine
minicab, dubbed Minicab No1 and sporting its unique registration
plate TAX 1 (more about that in Chapter 6), had been specially
craned up onto the roof of the building to become a backdrop to the
speeches and to the publicity photos.

As the taxi-cab section of the Transport and General Workers'
Union had turned the launch of the operation into a political issue,
no member of Her Majesty's Government was allowed to accept

an invitation to open the ceremony, so the invitation was issued and accepted by Monsieur Guitton, Chef du Cabinet, Ministère de l'Industrie et du Commerce (the French Minister of Trade), who duly performed this service, and, in his speech, he gave his views on the large order for Renaults, which were of particular significance in the light of the then current negotiations of the United Kingdom to participate in the European Common Market. The Minister poured a bottle of champagne over the bonnet of the Renault, whilst representatives of sponsors, including Air France, as well as of Welbeck Motors management, uniformed staff, and the gentlemen of the press looked on. The champagne flowed to toast a positive

Michael Gotla addresses the gentlemen of the press and invited guests, as the official launch of the minicab operation is due to start. Mr R S Walker is seated to the left and a young Graham Walker is standing behind. (Graham Walker – Welbeck Archive)

Monsieur Guitton, Chef du Cabinet, Ministère de l'Industrie
et du Commerce christens Minicab No1, whilst Mr R S
Walker looks on (Graham Walker – Welbeck Archive)

beginning to a new chapter of alternative transport for Londoners and tourists alike. Mr Walker also welcomed the guests and thanked the British press for the responsible manner, in which, he said, they had reported this challenging venture. Also he paid tribute to Mr Richard Sawrey-Cookson of Michael Clark-Hall and Associates, which had conducted their public relations since October 1960 and which had been responsible for the minicab project from its inception. In Michael's own speech, he commented that minicabs had had a good press, because they had always told the press what they were doing.

On the rooftop terrace at Taunton Place. 'Miss Minicab' Jane
Simpson raises a glass of champagne in front of Minicab
No 1 'TAX 1'. (Graham Walker – Welbeck Archive)

Michael enjoys the party atmosphere, much to the amusement
of guests. (Graham Walker – Welbeck Archive)

Welbeck Motors mechanics, suitably dressed in clean
white overalls, take a break to view the launch of the new
operation. (Graham Walker – Welbeck Archive)

One of the first minicabs will soon depart for operation. Note the Renault Dauphine on top of the waiting Welbeck Motors' car transporter, on which can be seen to the bottom right of the image, as photographed from the building above. (Graham Walker – Welbeck Archive)

The drivers and minicabs formed up in parade at 2.15pm. when the party itself was underway and, from the rooftop, the spectators could watch events below. But, by 3.00 pm, the building had been cleared, as it was at this time that the new fleet would officially commence operation. The initial cost of launching the scheme, including publicity, advertising for staff etc, (but not including the vehicles) was said to have cost in the region of £4,000, which was a not insignificant sum in 1961.

At Taunton Place, Mr Walker remained in the busy radio room for the rest of that day until three o'clock the following morning,

Michael is seen on the far left of this image, addressing his team of minicab drivers just before operations are due to begin. (Joel Frorath)

As the 3pm start of operations is about to commence, drivers are now beside their vehicles, so that they can set off in unison, whilst being filmed by the media. (Graham Walker – Welbeck Archive)

Michael Gotla, Minicab Man. This iconic image is believed to
be a personal favourite of Michael's and it captures a special
moment in time. (Graham Walker – Welbeck Archive)

BRIEFING : Drivers of the Welbeck Motors Mini-cab fleet, which began operations in London on Monday, are briefed by Mr. Michael Gotla (left foreground) near their garage. Mr. Gotla, the managing director, has introduced 200 Renault Dauphine Mini-cabs to the streets of London.

Another image taken of Michael briefing drivers just ahead
of the 3 pm launch of the minicab fleet onto London's
streets. (Graham Walker – Welbeck Archive)

describing the first twelve hours of operation as 'sheer hell'.

Interestingly, in a single story, which broke through World Press News on 14 July 1961, Michael Clark-Hall & Associates Ltd, the public relations company, which was behind much of the successful publicity campaign for the Welbeck Minicab venture, was reported to have resigned its account with Welbeck Motors. Mr Clark-Hall was quoted as telling Michael Gotla 'that the arrangement between the two companies is not as such as to be workable on a long term basis. I am sure you will agree that, from everybody's point-of-view, it would be better if your public relations were to be handled by another organisation'. A certain Cynthia M Arakie from Transworld News Service of London W9 wrote to Michael personally on that same day, in order to offer her organisation's services for Welbeck Motors as a new client. She made mention of the fact that, in a previous position as Chief Information Officer for the British Safety Council, she was the person responsible for launching the car safety belt campaign! It is not known if this offer was ever taken up.

Chapter 4

The Operation underway

After the previous months of regular publicity and newspaper articles, all eyes would now be on Welbeck Motors and how its minicab service would develop. Whatever happened now in daily operation, would surely be reported by the press.

All media eyes were on how the first day would unfold.

Within five minutes of the minicabs moving off news came of a minor clash. A vehicle driven by Mr Ronald Gay, a former long-distance lorry driver, who had been sent to 'The Planetarium' in Baker Street, was hemmed in by two taxis, until the police arrived and asked them to move.

One photographic image, which was captured by the press, was of world middle-weight boxing champion Paul Pender who, with his brother Bill Pender, a New York policeman, was staying at the Carlton Towers Hotel in London, ahead of a match with Terry Downes at Wembley. He was snapped as he stepped out of a Welbeck Minicab, having enjoyed a sightseeing tour on the first day of the

The first minicab fare for Vivien Batchelor was boxer, Mr Paul
Pender. ('Evening Standard' 19 June 1961 – Welbeck Archive)

venture with female driver, Vivien Batchelor behind the wheel, this
hire being, incidentally, her very first fare. He had called through for
a minicab and was quoted as saying 'He liked to try anything new
and that he had read a lot about the controversy; 'If there is a fight
I want to be in it,' he said.

After this, Vivien had little time to reflect on this event, as,
heading back to Taunton Place, her radio crackled again 'Blue 8,
Blue 8, (the car's call sign) are you near 'The Dorchester'? Her next
fare, who was to be taken to Finchley, was a smartly dressed matron,

A crowd of onlookers opposite the Taunton Place premises seem interested in the Austin FX4 Taxi and its fare paying passenger seen alongside a brand new Renault Dauphine minicab on the operations launch day (Getty Images)

clutching a Pekingese dog, who almost had to fight her way through the interested crowd, which had gathered upon the minicab's arrival. The lady commented that she felt that the advertisements on the vehicle looked a bit garish, but that she would use a minicab again, as it was so much cheaper than an ordinary taxi, and her dog hated buses. Vivien also received comments from taxi drivers, who shouted out from their cabs, 'Do you sell ice cream in that thing?' or 'You only want a few balloons and you would be a proper circus'.

Another female driver, 36 year-old Norma Bailey, in car 'Blue 48', was halted in Green Park by two policemen for the offence that minicabs displaying advertisements were not allowed to travel

Vivien Batchelor smiles for the camera, as she sets off to
pick up a fare. (Graham Walker – Welbeck Archive)

through the Royal Parks. Norma explained that the problem was
being sorted out by the appropriate authorities, but the policemen
decided to book her anyway, until it was sorted out. The passenger
on board was a Mr Geoffrey Parkhouse, who commented on this
incident in his column for 'The Daily Herald' newspaper. At the same
time, another minicab, 'Blue 66' was having trouble, complaining
that three cabbies had boxed him in. Minutes later another minicab
had been boxed in outside the Royal Academy.

Contrary to the huge amount of publicity explaining how the
minicabs would operate, the start of operations was somewhat

Lights, camera, action! The British Pathé newsreel camera captures brand new minicabs setting off with their first passengers from outside the Crawford Street Showrooms (British Pathé)

Police are called to a dispute in Fleet Street, London on
19 June 1961. (Graham Walker – Welbeck Archive)

different behind the scenes than that, which the Press reported. Former Welbeck Motors salesman, Hugh Ruding Bryan, recalled that about one hundred and fifty uniforms were issued ahead of the launch, which gives the impression that a similar number of Renaults were being pressed into service, which is not the two hundred, which was the figure widely printed in the Press. Of course, even the former figure was not accurate, as some of the minicabs were lost before even they went into service. Press publicity invited potential candidates to join the company: 'Be a £1,000 a year driver. Do you know your London? Do you like working on your own initiative and being paid by the results? Come and be one of our drivers. We need men and women with sociable, pleasant personalities and who won't mind driving all over the country at times. Pay while training. Bonuses. Promotion to staff positions for suitable candidates'.

On the busy London streets, a Renault Dauphine minicab
is closely accompanied by a traditional Austin FX3s and
an FX4. (Graham Walker – Welbeck Archive)

'WelbeckCabs2' by John 'Hoppy' Hopkins – 1961. (Estate of JVL Hopkins)

Applicants were plentiful. On a Saturday in May, Welbeck Motors had taken the front page advertisement space in 'The Evening Standard' calling for applicants for the job, and, on the following Monday, the company received in the region of seven thousand replies. Amusingly, it was noted that a mass application had been received from ALL of the van drivers from a large London department store! Indeed the pay was very fair, at a time when a taxi driver's wage was said to be £4 a week on average. In contrast a minicab driver could be expected to earn a minimum wage of £7 per week or one third of the takings, if this were more. Early predictions expected that up to £13 per week could be achieved.

Before the start of the operation, it was noted that only 10% of the applications had come from women; at the start, the actual number was fifteen and the management had hoped that it would have been more. Within the surviving newspaper articles from the Welbeck archive the names of not only Jane Simpson, Vivien Batchelor and Norma Bailey are reported but also of Kathleen Edwards, Betty Puri and Avril Strong are also mentioned.

Those persons, who had been in the Services, were engaged, but people from all walks of life applied, including bank managers, choreographers, delivery men and even taxi drivers!

The only trouble was that certain 'villains' had also applied and had been successful in their application. But they took the cars and never returned them; at least twenty-five cars never came back. Although the interviewing process was stringent, including the checking of driving licences, answering a barrage of questions on geography and being taken out for a driving test, some characters slipped through the net and applied for a job starting 1 August; once the school holidays were underway, they filled up the minicab

with petrol, packed the family in and went off on holiday in it! Eventually, cars were tracked down in Manchester, in Cardiff and even one in Paris! It was suggested that one car was never, ever found. One car disappeared for two months, before it was located in a remote village in Wales, where it had been used as a tractor to pull a plough during the day and then turned into a camper van at night, with all the seats removed and a mattress and camping stove installed. The driver was never found. After these episodes, it was decided quickly that all staff would have to be fingerprinted for security reasons and, with that, fifty staff left, as they would not allow themselves to be checked!

Although shrewd in the motor trade, Michael had not been prepared for this kind of start to the minicab venture. The only legal way for the business to operate was for the cars to have the bookings radioed through, then for drivers to collect the fare and to undertake the journey. If a car was hailed, the details would be passed on for another cab to collect the fare. Although it had been made clear to drivers that a passing cab being hailed by a member of the public was strictly forbidden (and Michael genuinely believed that the drivers would adhere to this rule), in practice the temptation for some drivers was too much. To be able to obtain a fare without a London taxi driver witnessing this happening, was too tempting. That is not to say, however, that all was not good with the staff. Many drivers were well suited to the role of minicab driver; one applicant, for example, had passed the test of the Institute of Advanced Motorists. Others were already well versed in their knowledge of London's streets from previous employment, which had gained them their own knowledge, albeit not in the same way that a London taxi driver had to prepare, in order to be licensed.

In the control room at Taunton Place, both telephonists and
controllers will be inundated as the afternoon of 19 June 1961
moves into the evening. (Graham Walker – Welbeck Archive)

Telephones ring in the Minicab control room. *PETER WAUGH*

Looking from the opposite direction, David Evans is seen
standing to the right inside the busy control room, as
telephonists take calls from the public. (Peter Waguh)

Michael had sent out a memo to all staff in order to make Welbeck
Motors' position unmistakably clear on the subject of 'plying for
hire'. The message read 'First, you take your jobs from the radio (in
the minicab) or you might be given an advance booking when you
come to work. YOU ARE NOT ALLOWED TO PLY FOR HIRE,
that is to say, pick up people in the street. But when someone hails
you, you will (if traffic conditions permit) stop and deal with his
order. Don't stop in heavy traffic. When you've stopped, if the client

David Evans busy at work on the launch day of the minicab operation
is filmed by the British Pathé news reel camera. (British Pathé)

wants a very short run just give him a card (which tells him all about Welbeck Minicabs) and tell him that the London taxi is the best for a short job. But if he wants to go a fair distance, tell him you will get a minicab to him very soon. Leave him – don't stay still by him – and as soon as you get out of sight, radio his order back to base. We will direct another minicab to him soon. If the whole system is overloaded and this procedure is unworkable, the radio control will have already told you – say – "10 minutes delay" or "20 minutes delay".

The control room was soon handling in the region of what was reported in Welbeck Motors' surviving internal paperwork to be forty thousand calls a day, with the office staff sitting in the room, seven in a line, each with three telephones and a pad, and a long conveyor belt running down the length of the room alongside them, on which the operators placed their slips of paper with the written telephone orders on. These would travel round to the dispatchers, who would pick them up and radio out the calls to the minicabs. For example, a dispatcher would put out a call for 'Trafalgar Square, Trafalgar Square.' over the radio. Eventually, a minicab driver would radio back with their number and say 'I am here at this location, what does the person hiring look like?'

If any calls appeared to be suspicious, the operator would raise his hand to ask a manager to assess it. The operators would soon gain experience on how to spot a dud call and they could tell by the way the caller would hesitate, when asked how to get to the address they had given. One woman was employed full time checking on suspicious calls, as she discovered that people did not appreciate being woken up in the middle of the night to be asked if they had ordered a minicab. Hoax calls were a constant problem; during one night alone, it was estimated that 90% of the calls taken were 'duff ones' probably instigated by cabmen, and that nine emergency calls had to be made to the police, asking for help in sorting out incidents.

Quickly, things became heated with the taxi drivers, and on one occasion it is said that a late night message from a minicab driver was relayed back to Dave Evans at the operational centre. It was in the early hours of the morning and a call came through from a driver saying 'It's blue 13, blue 13, I am surrounded by taxi

Jane Simpson is photographed and interviewed by reporter, Bill Evans for a special 'Evening News' feature on the first day of minicab operation. (The Evening News and Star – Welbeck Archive)

drivers. I have locked the car from inside and they are all around me rocking it from side to side. I am in Hyde Park what should I do?' Dave was renowned for his dry sense of humour and, possibly knowing what the inevitable outcome was going to be, he replied, 'Blue 13, repeat after me. . . "Our father, which art in Heaven, hallowed be thy name . . . !" Whereupon the taxi drivers turned the minicab over onto its roof with the unfortunate driver still trapped inside!

An interesting view filmed looking down into a local street captures busy minicabs loading passengers on the first day of operation. The passing Renault Dauphine seen at the top of the image is not a minicab! (British Pathé)

The servicing of the minicabs was undertaken by a team of mechanics on the third floor of the Taunton Place premises. This procedure necessitated the use of an internal lift in the building, into which the vehicles could be carefully squeezed, in order to elevate them to the higher level for the necessary maintenance work to be carried out. Due to the nature of the business and to Welbeck Motors' other activities, the workshops were in constant use throughout the twenty-four hour period, seven days a week. Teams worked shifts of either Monday to Saturday or Sunday to Friday between 08.00 and 18.00. There was a gap period shift between 18.00 to 22.00, when the night shifts came on from 22.00 till 08.00 in the morning, working on the basis of four men giving seven day coverage with three men per night. The Saturday and Sunday shifts started an hour later at 09.00, and there was also a breakdown service between 22.00–08.00 over a seven day period, manned by salaried staff.

One day, Michael, who was known for his occasional tempers had a huge row with a group of mechanics and he had told them all that, unless they arrived on time in the morning and sharpened up their attitude, he would deduct the cost of any damage to the cars from their wages and, furthermore, if they left early, he would sack the lot of them. That night, a group of these mechanics, who were each earning in the region of £15 a week, all left early, but not before they had opened the lift gates on the top floor and pushed three minicabs down the open lift shaft on top of each other. It took over three days of effort to haul the cars back up and into the building again, whereafter the lift needed repairs, and the cars, beyond repair, had to be sold for scrap.

Companies in the public transport sector invariably need a lost property office or storage area for such paraphernalia, and Welbeck's minicab operation was no exception. A reasonably sized room was allocated for this purpose, and, within six months of the operation starting, it became full of everything from suitcases, cameras, gloves, umbrellas, coats and other items which one would not imagine could possibly be left by a fare-paying passenger. The problem was heightened by the fact that the Renault Dauphine had a rear engine and its luggage boot was in the front, where, traditionally, most cars engines were fitted. A Renault Dauphine was not an obvious type of vehicle for minicab use, especially when compared to a traditional London taxi cab. For example, a passenger might be taken to Heathrow Airport and, upon arrival, he would hastily jump out of the car, eager to ensure that he was not late to catch a plane and quite forgetting that his possessions were still in the luggage compartment in the front of the car.

Normally, whilst in service, the minicabs would not remain back at Taunton Place for very long. At times, perhaps twenty or thirty cars would arrive for the short changeover period, but, usually, the cars would be back out on the roads again with their drivers quite quickly. Sometimes a minicab had been left parked in a nearby street and 'unknown persons' would remove the car and drive it to a location, which would not be easy for Welbeck Motors' staff to find. Occasionally, a minicab would have its two-way radio left to 'transmit,' thus blocking the whole radio channel system until either the car could be located or until the battery on the car had run itself flat. This became a recurring problem, as the technology was not as advanced as it is today.

On the 9 July, Michael put out a statement that, in effect, anyone wishing to hire a minicab would have to travel a minimum distance of five miles. He said that this step was being taken in order to reduce the number of customers, who were jamming the switchboards with incoming calls. Demand had been much greater than anticipated, with the project having gained much more publicity than it could ever have hoped for. It was oversold constantly and there was too much in the newspapers about it. With two hundred minicabs now on the road, it was estimated that sometimes, there was forty people wanting every cab. The situation was not helped by malicious calls blocking the lines. At times, it was estimated that there was in the region of 700 'dud' jobs every day, and this bottleneck was being increased by there not being enough administrative staff or 'dispatchers' with a deep enough knowledge of the streets of London, who could deal promptly with enquiries. Sometimes, the organisation was receiving up to two thousand calls per hour.

In the early months of operation, with many of the staff being inexperienced (and thus, in the eyes of the management, being inefficient), some had to be replaced. In fact, the same problems reappeared, because more bookings were being taken than the minicabs could handle, and, consequently, this gave rise to complaints from clients. However, the situation gradually improved as new staff were recruited.

It had been expected that fifteen thousand calls per day could be dealt with, but the actual number was forty thousand and, at times, the Welbeck Motors' telephone exchange was blocked with so many calls to the Welbeck telephone number. Statistics from the GPO showed that, on the second day of operation alone, eleven thousand people had unsuccessfully called for a minicab. Finding drivers was an increasing problem, and, due to the hostility of the London taxi drivers, retaining existing drivers became even more problematic. Being jeered at by the taxi drivers and being boxed in by their vehicles was commonplace and reports of minicab drivers being attacked, beaten up and threatened was a darker side of the story. Sometimes it was projected that two hundred vehicles would be out on the early morning shifts, but only one hundred drivers would turn up. Many of the original drivers had proved to be inefficient or lazy and, during the very hot spell of weather that year, absenteeism was rife, for various reasons. In fact, there were more than enough minicabs available for use, but there were simply not enough regular drivers to be found to drive them.

An example of a sacked driver's attitude was highlighted, when it was discovered that a thirty-three year old driver, who, as a former shift leader had been dismissed by Welbeck Motors, had decided to 'borrow' a minicab on the following day from the Taunton Place base. After stripping it of its' advertisements and after using it privately for two weeks, he was eventually arrested, after surveillance by the police in Hornsey. When taken to court, the man said that he did it out of revenge for being unfairly dismissed. Although previously of good character, the former employee was fined a total of £50, and was disqualified from driving for two years for using a vehicle without insurance.

Early on, orders for jobs of around ten miles or over only were being accepted for the following day, with advance bookings for long distance journeys being welcomed, this being in order to try to stabilise the availability of minicabs for the day-to-day operation. As far as payment for hire went, it was now only on a cash basis, due to the low-priced service. An additional phone number of HUNter 1250, running through a parallel system, was installed and was given out to regular clients, in an effort to ensure that they could get through without encountering many of the earlier mentioned problems. After the first eight weeks of operation, matters had started to settle down with something like two thousand jobs being undertaken daily, and in order to ensure that these jobs were being carried out efficiently, drivers were required to radio back the message 'customer contacted' every time a client was picked up on time. With this procedure in place, statistics showed that, in fact, approximately 95% of the jobs had been efficiently and correctly carried out. The remaining 5%, which still needed to be addressed, was made up mainly of other customers, who were missed or picked up late, and it was hoped that these operating problems could be corrected before the end of the year. Furthermore, it was hoped that, during the September, more minicabs would be put onto the road, in an effort to improve the service, for which there were currently too many customers and not enough vehicles.

One point, which Michael still stressed, was that the minicab was NOT in competition with the London taxi and that its operation had never pretended to be a high-class private hire service. He maintained that it was really a new kind of transport; use a taxi for short runs, but use the minicab for longer runs around the town or to the suburbs or the country. He even went on to say 'hire a minicab by the day to deliver your parcels – it's often cheaper than running a van'.

The opposition from the cabmen did not die away as reports had expected initially. On 29 June, over three thousand taxi drivers crowded into the Central Hall in Westminster to protest against alleged infringements of the law by minicab firms. Even though the speaker Mr Godfrey Stevens, vice chairman of the London Motor Cab trades joint committee, stated that Michael Gotla did not want a fight with them he went on to say, 'Mr Gotla is going to have a fight (loud cheers from the drivers). He is going to have a lot of trouble, until these abortions are driven off the road completely, or do the job they are entitled to do, which is legitimate private hire'. A three-point plan was decided upon at the meeting, which was to report to the police all cases of minicabs plying for hire, to consider a boycott of all goods advertised on the minicabs and lastly to press MPs to draft new by-laws to control minicabs. But the cheers inside the meeting turned to anger outside, as the cabmen left the meeting to find the police driving away several of the fifteen hundred taxis, which had been left parked around the building.

A reporter, who had been covering the story, had called for a minicab to collect him after the meeting had finished, and, after he watched the hundreds of taxis dispersing, he could not understand why his minicab never came for him . . .

As always, Michael, had an amazing eye for promoting the business and he was keen to acquire the telephone number Mincing Lane 4222. The reason was the method of dialing a telephone number in those days; letters were also present on the telephone dial, which would provide one with the correct exchange to be connected to. So, for example, should a caller want to phone for a minicab and he could not remember the company's number, by dialing 'M-I-N-I-C-A-B' (M-I-N-4-2-2-2) he would be put in contact with Welbeck Minicabs.

The Postmaster General, Mr Bevins of the Post Office Telephones, had refused to allocate this number to Welbeck Motors, but Mr Speir MP stepped in to ask the question why? The written response from Mr Bevins stated, 'The type of dial used in the London telephone numbering systems is not designed to be used for advertising and it would be impossible to provide numbers of this sort for all who might want them'.

As the month of July passed, there were reports of gangs targeting the night shift drivers. On several nights in succession, drivers had been dragged from their minicabs and beaten up. Usually, the attacks appeared to take place in the early hours of the morning. One example was at 03.50am, when a moving minicab driven by

Mr Albert Beeland was edged up onto the pavement in Vauxhall Bridge Road by a Jaguar, out of which four men jumped and he was brutally attacked and left in the gutter. The radio equipment was then wrecked, although the driver had just had enough time to radio back an emergency call, giving the number of the attackers' car. Neither had daytime incidents with taxi drivers lessened.

At London Airport, two taxi drivers had complained that a minicab driver was parked in the passenger channel, asking passengers if they had booked a minicab. The taxi men stated that this amounted to 'plying for hire', which, on the face of it, could be deemed so, because the pre-booked passenger had not, at that stage, come forward. The taxi men felt this was blatant stealing of their livelihood and just plain provocation. Trouble erupted and the minicab driver, Stephen Wojcik was punched and kicked, whilst astonished members of the public looked on. Other taxis quickly arrived from the airport's main taxi-rank and a squad of Aviation Constabulary Police was called. The minicab driver was too shaken up to be able to continue working and the police had to persuade him to leave his cab, while statements were being taken. Two taxi drivers were later charged with obstruction and others cautioned.

On television, the BBC invited licensed taxi drivers and their officials to take part in a debate alongside drivers and representatives of the owners and minicab officials. The interviewer, Peter Woods had an open discussion on television, where the frustration and tempers of the taxi drivers were aired, alongside the opposing views, which included representatives from Carline, Sylvester and Michael Gotla. One taxi driver said it was 'a diabolical liberty' that people, who had not been subject to any test at all, should be allowed to

Both minicab and taximan spot the photographer of this image
in London traffic. (Graham Walker – Welbeck Archive)

carry out the same work as qualified drivers, licensed by Scotland
Yard. Another criticised the uniform of the 'minicab' drivers saying
they looked like members of the Foreign Legion. Michael repeated
previous statements that minicabs were a public necessity and
provided a service for those, who did not normally use taxis. He
said that, if they were a success, he would ask for his vehicles to
be placed under Government control, similar to that of taxis. On
being questioned about the advertisements covering the larger part
of his vehicles, Michael's reply was that buses, which are public
passenger vehicles, also carried advertisements and that he could

see no difference between the one and the other. The subject of the 'Knowledge' came up and viewers were shown scenes at the headquarters of the London General Cab Co Ltd, where taxis, finishing their shifts, were being washed and overhauled after every turn of working.

Some of the many and varied stories in the newspapers were good news about the operation and, as the weeks passed, it showed how the media was still very much interested in the Welbeck Motors operation. A minicab was spotted in Kings Lynn, Norfolk and the driver interviewed by a reporter from 'The Lynn News & Advertiser'

'Mini-cab seen in Lynn'. (Lynn News & Advertiser – Welbeck Archive)

in late August. The driver, a Mr P R G Symonds from Gaywood, who had served in the RAF, was originally from the area and, upon driving a customer there, could not resist dropping back in to see family for a few hours, whilst he was resting.

The minicabs turned up in unexpected places when hired; for example 'a look of despair' described the taxi drivers' faces when they witnessed two pre-booked Renault Dauphines, covered with advertisements, arriving at Epsom Register Office to collect guests after a wedding!

Michael's earlier statements on what would become a 'test case' of their operating procedures was put to the test on 15 July, when a summons (by taxi driver, Edward Wall,) against Welbeck Motors' driver, Mrs Kathleen Edwards, was heard in front of a packed public gallery at Marlborough Street Court in London. It had been alleged by Mr Wall that the driver and owners of the said minicab had used an unlicensed Hackney Carriage in contravention of Section 7 of the Metropolitan Public Carriage Act 1869, by plying for hire in Upper Grosvenor Street, Mayfair on 20 June. It transpired that the taxi driver had observed from his vehicle a couple standing by the kerb, who were hailing hired cabs, which were passing by. A minicab, driven by Mrs Edwards, pulled into the kerb just short of the couple, she got out and spoke to them. She then walked back to the minicab with the couple, who got in. At this point Mr Wall got out of his taxi, went over and, upon opening the door of the minicab, he asked the couple inside if they had booked by 'phone or through a previous minicab. The reply from the couple was 'No, the driver is willing to take us where we want to go'. The police were called and taxis in the local area surrounded the minicab, so that it could not move. When Mrs Edwards was interviewed she told the constable 'I met one of our minicabs around the corner, the driver said that the couple had asked him to order one of our cabs for them. He had radioed control, but, when the couple got in, I was surrounded by cabs so I couldn't drive off'.

Mr Leo Gradwell, the Magistrate, fined Mrs Edwards £5 and Welbeck Motors (Minicabs) Ltd £5 with £26 5s costs. The magistrate had refused Welbeck Motors' counsel, Mr Bathurst-Norman, an adjournment; he had stated that it was a difficult case involving a lot of law and likely to be a test case. He went on to say that Welbeck Motors had unfortunately mislaid their summons and had been unable to instruct their solicitors promptly. It was embarrassing for him, as his client had instructed no further than for an adjournment and he did not know his clients' plea and was thus unable to examine the prosecution witness.

Michael did report later that he would be appealing against the decision, due to the refusal to grant an adjournment, which would have allowed the defence time to prepare. It was said that the alleged offence had happened only two days into the operation when 'everybody was in a confused state of mind 'and that Mrs Edwards had left the job after five days of her own accord. At first, the driver was going to plead guilty, but then changed her mind to not guilty, whereupon Welbeck Motors decided to request the adjournment.

Other cases would follow. An attractive blonde taxi driver's wife, a Mrs R Joseph, spotted a minicab outside the door of her home and decided to persuade the driver, Mr Dennis Goodwell, to give her a lift. She then reported him to a policeman for illegally plying for hire! Mr Goodwell had thought he 'was doing the woman a favour' and after pleading guilty in court was fined by £2 with Welbeck

Motors £1 plus £5 5s costs by the magistrate. He decided after eight days in the job that minicab driving was not for him, due to all the petty fights and bickering, and he decided to look for a job which was a little more peaceful.

Members of the public were generally in favour of the advent of minicabs; a Mr Ranger of Cloudesley Road, London, N1 had written to 'The Islington Gazette' on 21 July with the letter 'The Public wants more Minicabs', in which he described his personal experience, when he had recently ordered a minicab to go from his home to Waterloo Station with the family. For the three of them and their cases the fare was three shillings (3s 0d). On their return journey, they took a traditional taxi and the fare cost 6s. 3d. He said, 'Who wants to pay 6s. 3d. when you can pay 3s? It's common sense'. He went on to say, 'Just now, the public don't know how to get a minicab – I have been asked over and over again how I ordered mine'. He finished by saying 'What made me write this letter was the disgusting behaviour of the taxi drivers at Waterloo Station, when our minicab was driving away. They were tooting and booing like a lot of children. I'd like to say to Mr Michael Gotla: let's have hundreds of your minicabs, we will back you up – you won't lose'.

However, other businesses, were seeing the opportunities, which the minicabs could offer them. For example, The Bakers Arms Carpet Centre of Lea Bridge Road, Leyton started a scheme at the beginning of August, by which customers could travel to and from the carpet centre in a minicab free of charge. If prospective customers were unable to make their choice from the range of patterns available, they could be taken again, without charge to the carpet manufacturer's showrooms.

Michael's involvement in different circles was interesting; on 10 August he had joined the Society for Individual Freedom, the President of which was Lord Grantchester O.B.E, and which had an array of vice presidents including Viscounts, Sirs, Right Honourables and Lieutenant Commanders in office.

In the membership acceptance letter from the Honorary Organiser, Mrs Lilian Sutton, she made a comment about a report in 'The Times', to which she had, in fact, already responded ahead of Michael's membership being approved. This was due to the mixed press, which Michael was receiving at the time. Mrs Sutton had originally written to the papers two years earlier and had suggested that it would be in the public interest for minicabs to be allowed

A brand new Renault Dauphine in close company with
an Austin FX3 taxi. (Stanley Roth collection)

to ply for hire. She felt that it would be in the interest of free enterprise and competition, which in her opinion was healthy in any community, and that some of the incidents, which were taking place were wholly undemocratic. Also she had hoped, on behalf of the Society, that the Government would take note, and that initiative and enterprise was the way forward rather than restrictive legislation. Mrs Sutton felt that laws should be framed in such a way as to prevent unfair and unjust treatment of any individual. Tyranny is no less intolerable, because it is exercised by a majority.

By the end of August, there was a slowing down of trade within the motor industry and fairly regular meetings to assess how the Welbeck Motors organisation was operating became necessary, because large organisations can be more vulnerable to downturns than are smaller ones, due to their size. In a meeting on 30th of the month attended by the key men from various departments, including: Mr Kehoe, Mr Stanton, Commander C W M Milner, Mr E Fuller, Mr H Bryan (later Ruding-Bryan), Mr G Walker and Mr Drake-Briscoe, the key plan at the time was to ensure that sales cars, which had been put out as contract hire replacement cars, should be brought back in and sold off. It was advised that the second-hand car stock must be reduced quickly, and it was advised not to buy any more Renaults, except for replacement vehicles and to sell as many of them as they could. Even the possibility of selling the car transporter had to be considered. A focus on completing a deal to supply Pye with vehicles on contract hire was given priority to ensure delivery to this important customer.

From the surviving notes of a following meeting of the directors on 5 September 1961, Michael reported that the minicab project continued to make progress and that there had been considerable increase in the daily income. The taxi trade was still giving the company a lot of trouble, and expansion was handicapped by a shortage of suitable drivers. Steps had been taken to deal with the administrative problems by increasing the control staff, but it was emphasised that this increased cost must, as soon as possible, be spread over a greater number of units. It was agreed that expansion should proceed as rapidly as possible to three hundred effective units, at which stage there would be a further review. At the same meeting, Mr Finnie and Mr Emery representing the Corporation raised the subject that, since their last meeting, a notice had been served on the company by the Corporation limiting their financial facilities for all purposes and, in particular, for their contract hire fleet. The men felt that little or no attention had been paid to this directive, and no steps had been taken to prepare the company for the blow it must suffer by the cessation of the contract hire facility. This was bitterly resented by the working directors, who pointed out that the stock, the administration and the property position were all geared to a high throughput of vehicles on long lease, and that this position could not be reversed overnight, particularly in view of the present unsatisfactory state of the retail sales market following the 'little budget'. A fresh attempt was made to encourage the Corporation to continue to finance contract hire, but Mr Emery was adamant in his refusal. The boardroom argument continued forcibly, until Mr Finnie administered a closure by stating that the whole question of the Corporation's participation in Welbeck Motors would be brought under review in the very near future, and he undertook to give an early decision on the question of funding to secure a contract with the company, Pye, for some sixty eight vehicles.

Intriguingly, the next subject raised was a request by Michael for £50,000 to be made available for Welbeck Motors to pursue an offer by Renault Ltd, which concerned the acquisition of suitable storage premises and adequate spare parts' stock to run Renault's London distribution centre.

Mr Finnie said that he would place this matter before the parent board and would inform him of their decision. In the event, the Corporation did agree to the loan principle in October, but then decided that it would not proceed with the project at that time.

On 8 September, a confidential preliminary report was published for the internal Welbeck Motors management to see how the public relations side was running in the operation. Notes were made of the difficulties presented by frequent jamming of the telephone lines, with a particular emphasis on this tactic being employed as a disruptive weapon by taxi drivers to discredit the minicab service. The new alternative number would be further promoted to regular clients to avoid this nuisance. Attention was focused on developing the business market by targeting firms wishing to arrange travel for their executives, for which a special telephone number and VIP card were suggested.

It was known that taxi drivers often did good business at the horse racing and greyhound meetings, so the idea was considered of developing this market, using a special sports number. As minicabs were forbidden to have their own rank or rank facilities at airports, the idea of a sub-station was suggested, in order to solve this problem.

Further advertising, direct mail and editorial coverage was considered as an effective and long term boost to sales and as an inducement to passengers arriving at London airports to use the minicabs.

The use of teleprinters to link London with provincial cities also needed serious investigation, as this ploy was considered by the company to be the most direct and speedy method of communication. The plan would be to install teleprinters into small offices in selected provincial airport cities where passengers, especially businessmen flying into London, could simply telephone the local minicab office in order to book a minicab for a pick-up at (say) London Airport[*] or London Gatwick airports. This information would then be teleprinted direct to London bookings. On return journeys, London would teleprint the same information to the minicab provincial headquarters.

To encourage driving staff to take extra pride in their work, to become more efficient and to increase their passenger bookings, a 'Driver of the Week' incentive scheme was suggested, whereby an employee would be selected on the basis of the most passenger miles he had achieved. Drivers would also be encouraged to build up their own clientele list and to take the police test set by the Institute of Advanced Motorists. A further suggestion was a tie up with all the London theatre box offices, whereby their ticket envelopes carried minicab advertisements on the back. Finally, the seven page document noted how Welbeck Motors should be careful how it handled and managed the press by maintaining good press relations, but only as sanctioned by the directors.

[*] London Airport was renamed Heathrow in November 1968.

Chapter 5

Advertising on
the Minicabs

Michael's novel idea was not without some difficulties in minicab operation, due to laws in force concerning how advertising must be displayed in public places; however this source of revenue was crucial to the viability and success of the venture.

With his talent for advertising, Michael came up with a real first idea on how to offset some of the heavy costs involved in ensuring the success of the venture. This was done by selling advertising space on any or all over the various parts of the minicabs themselves. A company by the name of Gotham Advertising formed a new company called Minicab Advertising Ltd, which had specifically contracted to buy all the advertising space on these new minicabs at £70 per vehicle, per year.

A bespoke advertising brochure was produced with images of the minicab from various angles to tempt the prospective advertiser to

Image (Welbeck Archive)

Image (Welbeck Archive)

book a specific size of slot on the car, which would suit the company's style and budget. The introduction in the five page pamphlet explained that the minicabs provided a new kind of transport at a new sort of price, midway between public transport and traditional London taxis. It went on to say that minicabs were the first major transport innovation in London for fifty years and that they had been launched after tremendous planning and organisation, and had won a ready acceptance from the public. It stated that the advertising in this new manner would provide two benefits: the first was that minicabs would become generally recognisable by the advertising, which appeared on them, and, secondly, that advertising on this form of transport was new, sound and economical. The cabs would operate in the metropolitan area and market research had shown that the fleet would carry in the region of 15,000 passengers a day, but that the vehicles would be seen and noted by many millions more. A location on a minicab would give the advertiser a travelling advertisement around a densely populated area to tie up with its general advertising and also to feature special campaigns requiring a special form of media. "Remember! When they look for a taxi, they look for your advertisement" it said.

Advertising could be hired for a minimum of six weeks, then between periods of seven to thirteen weeks, fourteen to twenty-five weeks, with the final periods being for six months and for a whole year. The rates were as follows, on the various positions on the car, the cost being in pounds, shillings and pence (of course), and size of the advertisement in feet (') and inches (").

Position	Size	1 year	6 mnths	14–25 wks (pw)	7–13 wks (pw)	6 wks (pw) (min)
Nearside/ Off Side Panel*	7' 5" x 12"	£31.5.0	£17.5.0	13/9	16/0	18/6
Front Targets (per pair)	15"x10"	£6.5.0	£3.10.0	2/10	3/4	3/9

* The near-side and offside panel adverts ran across the front and rear doors and onto the rear side panels, which were considered to be the most visually prominent positions for adverts carried on the vehicles.

Position	Size	1 year	6 mnths	14–25 wks (pw)	7–13 wks (pw)	6 wks (pw) (min)
Back Targets	11"x6"	£6.5.0	£3.10.0	2/10	3/4	3/9
Front/Rear Doors	15"x6"	£2.15.0	£1.15.0	1/4	1/6	1/9
Side Pillars (per pair)	10"x2"	£2.15.0	£1.15.0	1/4	1/6	1/9
Top Streamer	34"x5½"	£6.0.0	£3.10.0	3/4	4/0	4/9
Back Panel	15"x20"	£6.10.0	£3.15.0	3/0	3/8	4/6

WELBECK MINICABS

On June 19th, 1961, Welbeck Minicabs came to London, providing a new kind of transport at a new sort of price midway between Public transport and traditional London taxis.

This is the first major transport innovation in London for 50 years and has been launched by tremendous planning, organisation and a ready acceptance of public interest.

The Minicab fleet comprises 500 Renault Dauphines fitted out perfectly as a modern form of transport. They are equipped with the most up to date VHF radio in Europe and are driven by first-class uniformed drivers. The cost of this tremendous venture has been £500,000 aimed at giving London a reliable, efficient and economical form of conveyance. While these points are of course of general interest, how can Minicabs help to publicise and sell your products - the answer to this is that every unit of this fleet will carry advertising on the outside together with certain panels inside. This will provide two uses, the first is that Minicabs will become generally recognisable by the advertising that appears on them, the second is that advertising on this form of transport is new, sound and economical. The cabs will operate in the Metropolitan area and Market Research has stated they will carry in the region of 15,000 passengers a day but will be seen and noted by many millions more.

A location on a Minicab will give you a travelling advertisement around a densely populated area to tie up with your general advertising and also feature special campaigns requiring a special form of media. We hope that this brochure will give you the cost and information you require; however, we are always at your service to assist you in any way we can.

Remember ! when they look for a taxi, they look for your advertisement.

MINICAB ADVERTISING LIMITED   HAYLER HOUSE · RONALDS ROAD · LONDON N.5
175 OXFORD ROAD · MANCHESTER 13.

Image (Welbeck Archive)

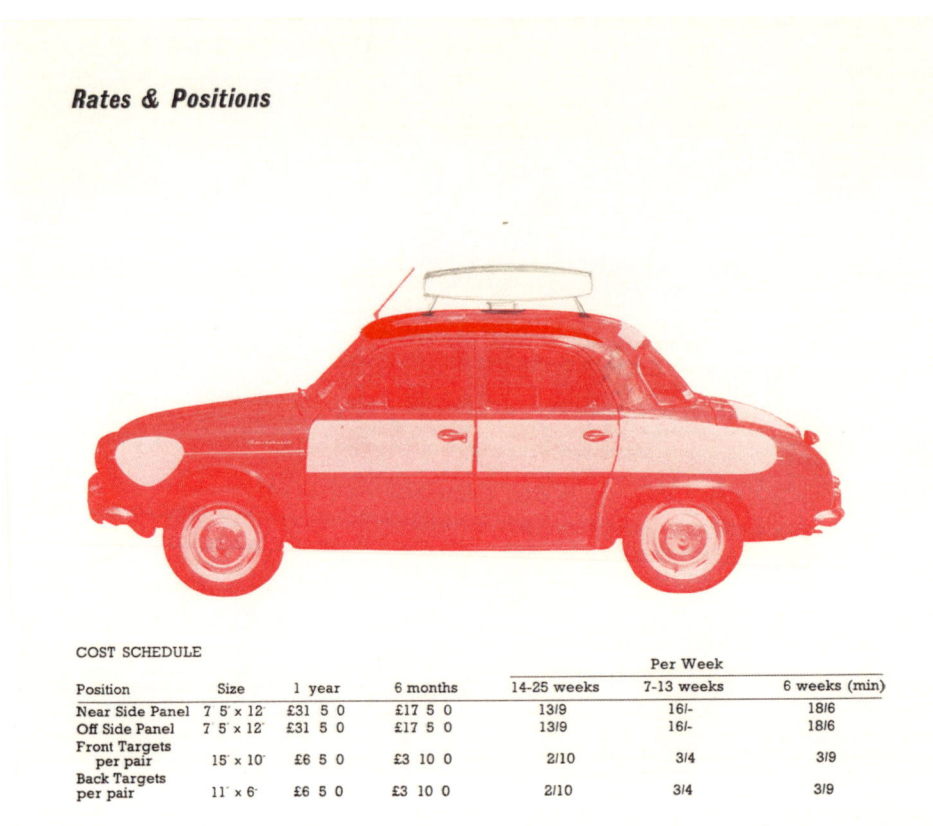

Rates & Positions

COST SCHEDULE

Position	Size	1 year	6 months	Per Week		
				14-25 weeks	7-13 weeks	6 weeks (min)
Near Side Panel	7′ 5″ x 12″	£31 5 0	£17 5 0	13/9	16/-	18/6
Off Side Panel	7′ 5″ x 12″	£31 5 0	£17 5 0	13/9	16/-	18/6
Front Targets per pair	15″ x 10″	£6 5 0	£3 10 0	2/10	3/4	3/9
Back Targets per pair	11″ x 6″	£6 5 0	£3 10 0	2/10	3/4	3/9

Image (Welbeck Archive)

Surviving publicity material shows that there was a proposal for a billboard (called a 'top streamer' in the advertising), which could be fixed across the roof of the car from front to rear windscreen, which would have offered an extremely prominent advertisement space, had it ever been introduced. Suffice it to say, that the advertisements on the minicabs would help to strengthen their visual impact on the streets, when in operation.

Counted among the various businesses, which took up the early offer of advertising, were Air France, Cyril Lord Carpets, Idris

Rates & Positions

Interior

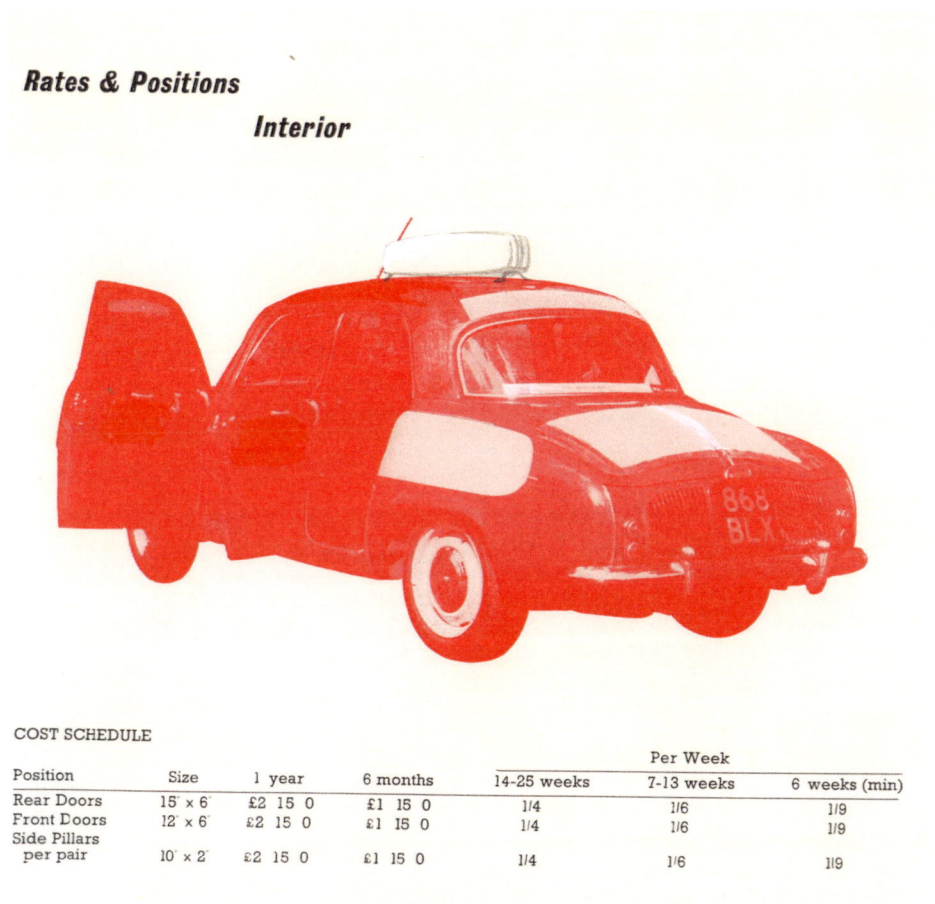

COST SCHEDULE

Position	Size	1 year	6 months	Per Week		
				14-25 weeks	7-13 weeks	6 weeks (min)
Rear Doors	15' x 6'	£2 15 0	£1 15 0	1/4	1/6	1/9
Front Doors	12' x 6'	£2 15 0	£1 15 0	1/4	1/6	1/9
Side Pillars per pair	10' x 2'	£2 15 0	£1 15 0	1/4	1/6	1/9

Image (Welbeck Archive)

Drinks, North Thames Gas, Britax seats, Pac-a-Mac and Wilhelm Cigars. As it turned out, this leap of faith on their part provided those companies with more media coverage and product exposure than they could possibly have dreamed of. The Welbeck Minicab name with telephone (Welbeck 4440) were predominantly displayed in a long landscape style on the side roof panels and above the front and rear windscreens.

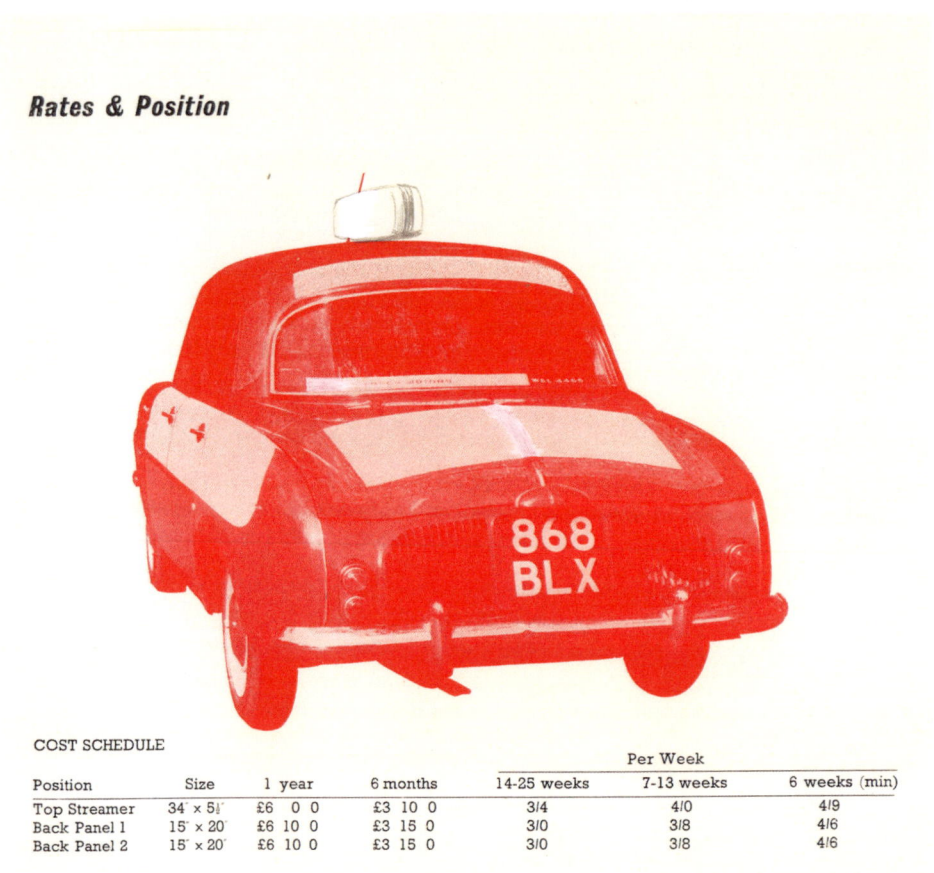

Rates & Position

COST SCHEDULE

					Per Week	
Position	Size	1 year	6 months	14-25 weeks	7-13 weeks	6 weeks (min)
Top Streamer	34' × 5½'	£6 0 0	£3 10 0	3/4	4/0	4/9
Back Panel 1	15' × 20'	£6 10 0	£3 15 0	3/0	3/8	4/6
Back Panel 2	15' × 20'	£6 10 0	£3 15 0	3/0	3/8	4/6

Image (Welbeck Archive)

Aside from this ploy being a sensation at the time, as Michael was the first man to have implemented this, it also raised a huge amount of revenue. The potential earnings of £35,000 per year, based on what could be as many as five hundred minicabs carrying advertising, was pushing the taxi drivers and the unions, frustrated by the venture, to the very brink of a strike, unless action was taken to remove Welbeck and their advertising from the road.

Image (Welbeck Archive)

Questions were even being asked in the House of Commons about the travelling advertisements. A question was put to the Minister of Housing and Local Government by Dr Barnett Stross, asking if he would take steps to control the placing of advertisements on cars and on taxicabs plying for hire, to which the Parliamentary Secretary to the Minister of Housing, Sir Keith Joseph replied 'No'!

The same question was also raised at a meeting of the London County Council, where it was put to the chairman, Mr Reginald

Stamp, that the council had the powers to check the display of advertisements on minicabs for producing an unsightly and undesirable effect in London's streets, to which the reply was that the council had no power to curb the display of such advertisements and that the responsibility lay with the Commissioner of Police.

Before he had even started using the minicabs wholly or mainly for the purposes of advertisements, Michael had already fallen foul of the law by organising a photo-shoot of a number of minicabs processing along the streets of London ahead of the minicab launch, his solicitors, Amery-Parkes & Co, had identified the regulation, which the police were considering using for a prosecution under the London (Waiting and Loading) (Restriction) Regulations 1958, which formed part of the London Traffic Act 1924 covering the restriction on waiting and loading in specified streets. There was a small paragraph under Section 14, which stated that: no person shall, in any restricted street, or in any other street, which is a thoroughfare, in the area contained within a circle of three miles radius the centre of which is the site, at the date of these regulations, of the King Charles statue at Charing Cross, either wholly or mainly for the purposes of advertisement:

Wear, or cause to be worn, any fancy dress or other costume; or Ride, drive, conduct, use or employ any animal or vehicle of any kind.'

Amery-Parkes were confident that, although the photo-shoot was for publicity purposes, once the minicabs were in service carrying passengers, there would be little chance, under this regulation, of a charge standing much chance of success. Scotland Yard had been contacted by them, and although the police would not commit themselves, they were inclined to take the same view at that time.

The sight of the Renaults in service brought comment and, sometimes, criticism from the public. In the many articles, which were being printed at the time, there were occasional references, such as 'dislike of minicabs bearing a rash of advertisements' and 'they look like mobile hoardings at a fairground' from some members of the public.

Chapter 6

That Registration Number!

In his uncanny luck of always finding further publicity, Michael could not have foreseen the opportunity of finding a unique and perhaps ultimate piece of advertising, in the form of a registration plate.

One day, a gentleman walked into the Crawford Street showroom announcing to staff, 'I'd like to sell you my car'. Salesman, Hugh Ruding Bryan looked out of the window and made a note of a Standard Vanguard parked outside. He commented, 'Thanks, old chap, but we don't really buy cars like that'. To which the gentleman replied, 'I think you ought to take a closer look first'. He was asking £200 for the car, but, once the laughter in the showroom had died down, they went outside and, to their astonishment it bore the registration number 'TAX 1' on it! Michael Gotla had also seen it and said to Hugh, 'We've simply got to buy this car.' Hugh said,

A minicab is captured on camera with the newspaper caption 'It seemed more than coincidental and a trifle cheeky to find this minicab with the registration number TAX-1'. ('Harrow Observer' and 'Gazette' 21 September 1961 – Welbeck Archive)

'But he wants £200 for it, with the number plate on'. Even though this was a considerable figure at the time a cash price was agreed on eventually, plus the gentleman's train fare back to Cardiff in Wales and a ploughman's lunch before he departed. Hugh felt that the Vanguard probably would not have passed another MOT without work. However the car was sold on and, in due course, the number was eventually transferred onto a Renault Dauphine minicab, a move which upset numerous taxi drivers even more.

Newspaper articles at the time differed in their reports as to how Michael came by the number plate. 'The Evening Standard' wrote on 24 February 1961 that 'A few days ago Mr Gotla saw a woman drive by in a car bearing the index number TAX 1; in half an hour he had bought the car from her and got the London County Council (LCC) right away to use the number for his No1 cab'.

This registration adorned a brand new Renault Dauphine minicab, which, for publicity purposes, had been craned up onto the roof garden of Taunton Place for the launch party of the minicab operation on 19 June 1961 and it was later spotted and photographed by the press on various occasions when in use. In late September, a keen reporter from 'The Harrow Observer & Gazette' had spotted a Renault Dauphine minicab parked near the Queensbury Roundabout at Honeypot Lane. The papers image clearly showed that the car was sporting the registration number TAX 1, which, as

Publicity shot of the rear of TAX 1. (Graham Walker – Welbeck Archive)

was noted, 'seemed more than coincidental – and a trifle cheeky – to find a minicab with this registration on'. Apparently, the driver confirmed that, despite displaying such a prominent number on his minicab, he had not had any trouble from taxi drivers.

After Hugh left Welbeck Motors, he managed to negotiate to retain the number plate, which he transferred onto a new Ford Consul convertible, sporting twin aerials and whitewall tyres. But, a year later, when he was working for another company and in a moment of weakness, he sold the registration on to a large taxi company in Basingstoke. It was then transferred onto a dark blue Ford Cortina, which was part of a fleet of identical cars at that time. The story does not end there, because, after six months, Hugh had received a call from a finance company to say 'Those scallywags in Basingstoke are not paying their dues, their finance is out of date and we would like you to repossess all their cars, they have to give them up'. Accordingly, Hugh went to arrange the collection of all these Ford Cortinas and, when he arrived there, he was moving between the rows of vehicles and found the car bearing the number TAX 1! Hugh re-acquired the number for nothing and after this it was transferred onto a succession of his own personal vehicles, including Range Rovers, an Audi, a Ford and a Volkswagen. In fact, the first new Range Rover, which Hugh purchased, was from Guy Salmon, who had been great friends with Michael Gotla over the years, always trading in cars, especially Armstrong Siddeleys. It was only a few years ago that Hugh finally decided to part with the number plate, when an acceptable offer had been put forward and, this time, it was finally sold. As a footnote, the registration number was originally issued by the County Council of Monmouth and dated from March 1958.

The registration number TAX 1 is seen at a much later date on one
of Hugh Ruding Bryan's Range Rover cars. (Hugh Ruding Bryan)

In February 2018 the registration TAX 1 found its way briefly
into the limelight again when it was put up for auction by Humbert
& Ellis Auctioneers of Whittlebury in Northamptonshire after
some forty years in private ownership, this time selling under the
auctioneers hammer for a respectable £92,000. No mention was
made of its previous brief flirtation with fame on the streets of
London some six decades before.

Life International Magazine 31 July 1961 – 'Editor's
Choice' Photograph (Welbeck Archive)

Chapter 7

Images of the time

In 1961, it was not hard to find images of the minicabs being drawn into the contemporary media culture something which is not always directly associated with the transport industry. Nonetheless, it offered unintentional advertising for the Welbeck Minicab operation.

The world was certainly a changing place at the dawn of the 1960s and the sight of these bright, colourful minicabs taking to the streets of the city obviously held a certain appeal to photographers for features in many of the popular newspapers and magazines of the day. A few images have been included here from the Welbeck archive, just to illustrate that at this small moment in time, the venture had captured something within the country's culture.

Vogue Magazine – 1 September 1961 (Welbeck Archive)

Tips for panicky cabbies

BY MARY MACPHERSON

Traditional taxi-cabs have enjoyed a complacent and unrivalled sway over the roads of London since horses changed to horsepower. This week though, they face competition. Mini-cabs are storming the streets with new and potent weapons—a flat rate of 1/- a mile with no extras, elegantly uniformed drivers, attractively small and manoeuvrable cars. The old-type cabbies have up to now been able to display the same take-it-or-leave-it attitude as the Post Office or The Only Man in London who can get you tickets for "Beyond The Fringe." The Londoner's feeling for taxis tends, therefore, to be mixed. Nothing can surpass the wave of deep affection that sweeps over us as an empty cab looms up in a rainswept street; but nothing can turn that affection into ashes quicker than the words "I'm on my way home, miss—take you towards Euston, if you like." And while we are breathless with admiration at the dashing way our driver is getting us to the theatre on time, the man in the private car next to us is almost inevitably winding down his window in white-faced rage. The majority of taxi-drivers seem to be calmly (or angrily, according to the mood you catch them in) unworried by the Mini-cab Menace. "They don't know their way around like we do," they say complacently. But in case their complacency is ruffled, and they feel their service could be improved. we asked some experienced taxi-observers for tips. And no taxi-driver has ever objected to those.

Take a tip from ✢ BARBARA GOALEN

"As a driver I would like them to be a little quicker when picking up fares in the middle of the street; it's so annoying when they block the road for ages. But I like to see them on the streets"

Take a tip from ✢ OSBERT LANCASTER

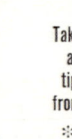

"Present day taxis were designed for enormous Victorian families and their luggage—nowadays there is usually only one person travelling at a time. What we need is a small cheap cab that you can throw away after a couple of years"

Take a tip from ✢ ANNA MASSEY

"I must say the vast majority of taxi-drivers are wonderfully polite. But there are a few who are so rude you could die. They could well improve their manners"

Take a tip from ✢ STIRLING MOSS

"They could bring their charges down—for the short distances you travel I don't think any comfort they may offer is a particular advantage. As a passenger I think the drivers are very good"

Take a tip from ✢ MARY QUANT

"They should really hand out false fingernails when you leave the cab . . . one always seems to break one's nails wrestling with the door. Failing that, they could well open the door themselves for you"

Take a tip from ✢ MUIR AND NORDEN

"They should put hassocks on the floor. Taxi-drivers always insist on talking to us, which means we have to kneel uncomfortably on a piece of coconut matting straining to listen through that little window"

Take a tip from ✢ JANE GASKELL

"They should go slower, ask if you want to get somewhere quickly, and if not, take you on a scenic drive. And they should spray the car with scent if the last passenger has been smoking"

Take a tip from ✢ ROY BROOKS

"A lot of taxis already have short-wave radios— we should be able to use them to telephone through to our next appointment. Being a person with no sense of time I'd find it useful to ring up and say I'd only be 10 minutes late."

Take a tip from ✢ MRS. MATTLI

"They should go faster. On the whole I'm a great fan of London taxis, but they are very slow compared with French cabs. I take a taxi to work every morning with my dog, and they seem to creep around"

Tips for panicky cabbies. ('The Tatler & Bystander')
21 June 1961 (Welbeck Archive)

'Cab Drivers' by John 'Hoppy' Hopkins – 1961 (Estate of JVL Hopkins)

Weekend Magazine. Womens Section 16 – 20 August 1961 (Welbeck Archive)

The Tatler & Bystander. (Front Cover) 21 June 1961 (Welbeck Archive)

The Cab War—5

Brockbank

" *I had my hand out, of course, preparatory to moving out into the stream and Joe here was coming in with the old 'rabbit's ear' winking away, to drop his fare—as no doubt Harry at the back there can bear witness to—when lo and behold . . .* "

" Booked a Mini-cab, lady ? Ain't seen no Mini-cabs around 'ere . . . but if you wants a REAL cab, lady . . . ! "

'I see what you mean. I'm not supposed to ply for hire, either.'

Newspaper cartoons. (above and left) (Welbeck Archive)

"So I said to myself, 'Sid, if you can't beat 'em – join 'em".
Evening Standard (Welbeck Archive)

Chapter 8

Minicabs in the Royal Parks

At the time, British law still dictated that there were public places where certain forms of advertising were not permitted. With the Welbeck Minicabs driving around adorned with mobile advertisements, they would inevitability be caught up in legal debates through their obvious street presence.

It did not take long for the question of breaking other regulations was brought to bear on the operation of the minicabs. The use of vehicles displaying advertising in the City of London did carry certain restrictions; for example it was forbidden to use a vehicle within a three mile radius of Charing Cross Station, if it was for the use mainly or wholly of advertising. Michael had researched the law and, after seeking the advice of the police (who agreed) determined that it was perfectly legal for Welbeck Motors to be able to operate in that area, as they were operating, as he put it, 'a taxi service'.

However, the matter of the Royal Parks was quite a different kettle of fish to interpret. Michael's interpretation of the law was this: 'You must not drive or use, without the minister's consent, a vehicle constructed or adapted for the purpose of any trade or business'. He felt that his minicabs fell within this regulation, but he was, as he put it, 'Hammering this out with the Commissioner of Police and within days the position will be made clear'. 'For the moment,' he told drivers 'use the parks and, if you are stopped, just tell the police that you have been told by the company to use them and that we are dealing with the Commissioner. Our lawyers are convinced that we are right in this temporary controversy'.

The National Archives contain some correspondence between Welbeck Motors' solicitors, Amery Parks & Co (who were also legal advisers to the Automobile Association) and Government Offices regarding the question of contraventions of the regulations by the minicabs using the Royal Parks, due to their all-over advertising being on display. If nothing else, the surviving letters have been included in this book to demonstrate how the wheels of decision-making moved slowly, with regular correspondence between Government Officers, before the decision of an eventual outcome.

The first recorded letter stated, that on 29 June 1961, a Mr R R Cole, Assistant Treasury Solicitor, wrote internally stating that it would appear Welbeck's Minicabs did constitute a contravention of the regulations forbidding the exhibition of advertisements, unless the Minister's prior consent had been obtained. He went on further to make reference to the Hyde Park Regulations 1955, the St James' Park Regulations, the Green Park Regulations 1955 and the Regent's Park Regulations 1932. Mentioning a previous discussion, Mr Cole understood that certain other activities, which were believed to be being carried out by the minicabs, might also constitute a further breach of regulations, although he could not be more definite without more detailed information. Mr Cole does mention the fact that Amery Parks & Co.'s previous letter to him did constitute an application for consents to exhibit the advertisements without prejudice against the true construction of the regulations. His comment was that the Government's answer to the application would be of course, a matter of policy.

The Advertisers' Weekly carried a small article about how, at both the London County Council and at the House of Commons, questions were being asked as to whether or not there was any power to check the display of advertisement and in each case, the answer was that there was no legal prohibition. In advertising circles, such questions had been raised repeatedly; by both advertisers and advertising agents, who were considering the idea of using minicabs for this purpose. It was stated that, under the Control of Advertisements Regulations contained within the Town & Country Planning Act, 1947, advertisements, which are displayed on or in a vehicle, are exempt from control. As a vehicle was a moving object, whether on a highway, railway or inland waterway, a bus, delivery van, canal barge or taxi cab could thus display advertisements, but could not be 'dressed up' as mobile advertising units. Such vehicles would contravene the regulation if they did nothing else *but* carry advertisements. Because it is employed as a moving vehicle, a minicab could not legally display advertisements, as could a mobile shop.

However, in 'The Times' of 2 October 1961, an article entitled 'Park Regulations broken by Mini Cabs' was published. Here, nineteen minicab drivers were summoned before Mr Leo

Gradwell at Marlborough Street Magistrates Court for exhibiting advertisements without the written permission of the Ministry of Works. There were twenty-two cases brought forward for using the west, south and north carriageways of Hyde Park between 17 and 23 of June that year, with the charge that Welbeck Motors were aiding and abetting. Interestingly, Mr Gradwell held that the cases were proven, but added that he had a certain sympathy with the defendants in the matter. He discharged the nineteen drivers and Welbeck Motors absolutely, but ordered them to pay ten guineas costs.

Although Mr Christopher Oddie for the defence had applied at the outset for an adjournment, saying that the extremely complicated legal issues in the case required further study, this application was refused on the basis that the magistrate could see no reason for the adjournment. Counsel argued that there was no reference to vehicles in the regulation concerned and that the carrying of an advertisement on a vehicle was not something that a minister could prevent. Furthermore, there was no molestation of persons using the park or annoyance to them, which was what had been envisaged in the preamble to the original Parks Act of 1879. Mr Oddie also contended that, since the minicabs were using the park only to get from one side to the other, they were not breaking the regulation, which was clearly intended to apply only to the persons, who entered the park for the specific purpose of advertising. The magistrate said that clearly nobody could possibly say that the advertisements outside the 'Palladium Theatre' were an annoyance. He was delighted to see them every day on his way to court, but he would be furious if the 'Palladium Theatre' suddenly advertised in his garden.

As the year turned into 1962, the situation rumbled on, this time with a letter from Mr Ogle-Skan of the Ministry of Works writing to E Ridley of the Treasury Solicitors Office on 30 January, which made the point, that, in previous correspondence, they had been told that 'minicabs' would be excluded from the Royal Parks, as long as the vehicles continued to use advertising. However, as the minicab operators had decided to test the validity of their exclusion with a batch of prosecutions and fines, resulting in the magistrates finding the offence proven, but only imposing a token fine, the Minister now felt that it would be unreasonable to continue to exclude the cabs from the parks and was considering issuing the necessary written permission to allow them in. A reply, drafted on 2 February stating that, in principle, there were no legal objections to Welbeck Motors having a permit (in suitable terms), subject to it being revocable at any time. However, when the official response was sent on 5 February, Mr Ridley's reply on the subject was lengthier. His letter started by saying that, before drafting documents, it was necessary to be quite clear what the Ministry was trying to do. The regulations across the Parks were not uniform, which meant it would not be possible to include everything in a single document. His department's research had shown that minicabs *per se* were not excluded from the Parks. The possible contravention of the rules was based on the fact that the vehicles were displaying advertisements. If the advertisements were to be removed, then there would be no reason why the minicabs could not use the Park roads.

Concern was voiced about defining or restricting the advertisements, and indeed how the Minister could give written permission for something, which the regulations prohibited, was for Mr Ogle-Skan to consider.

On the subject of what could be permitted, of the definition of the terms 'minicabs', 'operated' and 'Welbeck Motors Ltd', special note was made: the first two words did not have any particular significance, but the third was very precise. As the publicity about the new venture was at its height, the comment was made that Welbeck Motors had created such a visual impact with their minicabs carrying all-over advertising, that it had sown the seed of doubt in the public's mind as to what was the definition of a minicab. For most people, it meant that if one rang WELbeck 4440 and asked for a minicab by that name, a car with a driver would arrive in due course and it was a probability, if not a certainty, that the car would be a bright red Renault, 'defaced' with advertisements. Bearing this in mind, Mr Ridley was uncomfortable with the word 'minicab' being used at all. He concluded by questioning the reason for allowing Welbeck Motors to have a franchise to display advertisements in the Parks and suggested that it would not be unreasonable for others then to do the same.

On 8 February 1962, Mr Ogle-Skan replied, stating that what was required was to allow minicabs to use the Park carriageways to travel from one end to the other with or without their passengers. The reason for dealing with Welbeck Motors alone was that it was only that company, which had come forward to ask for such permission. By now, the Ministry of Works were aware that Welbeck Motors' vehicles had removed all the advertisements except those, which gave the name of the company, the telephone number and the word 'minicab'.

The situation had come to a head when Welbeck Motors' solicitors had written to the Minister pointing out that it had been nearly eight months since they had asked for such permission.

Mr Ogle-Skan asked if Mr Ridely could approach the solicitor in order to determine whether anything could be done to elicit suitable permission fairly quickly, for which action he would be grateful.

Finally, on 12 February 1962, Mr Ridley replied and, this time, he stated, 'I very much doubt if there is any occasion to do anything of legal significance'. As the minicabs were by now no longer displaying advertising, except for their company name they were no different from any other trade-marked vehicle, which would be using the Park roads. Whether the company name, number and the word 'minicab' constituted advertising, only a Court of Law could decide, he concluded.

In due course the Minister, produced a draft a document, which authorised the passage of 'minicabs' operated by Welbeck Motors through the Royal Parks, subject to the following conditions: passage would be restricted to roads between Hyde Park & Kensington Gardens; St James's & Green Parks; Regents Park; Richmond Park; Bushey Park and Greenwich Park. Vehicles would not be permitted to ply for hire or to stand or to park anywhere within the Parks. The hiring of a taxi was forbidden in the Royal Parks.

An incident reported in 'The Steering Wheel', the taxi industry's magazine in April 1961, told of a taxi, which, on setting down a passenger in Regents Park, was immediately hired by a woman and her eighty-four-year-old mother, who was recovering from an illness and had been resting in the Park, before deciding to go home. Upon the women entering the said taxi, a policeman and a policewoman came up and, in official language, pointed out that hiring a taxi in the Park was not permitted. The couple were required to get out of the taxi and the old lady was obliged to walk a considerable distance until she reached a street, where she was then able to hire the cab legally.

As a footnote, minicabs were not the only vehicles being monitored through the Royal Parks at this time. In June 1961, a South London clergyman was driving his shooting brake car, which carried stickers for the National Association of Boys Clubs 'Clubs Week', when he was ignominiously stopped and sent back out of a Royal Park, in which he had been driving!

Chapter 9

The Renault Dauphines

Welbeck Motors' choice of French cars to use for their minicab operation brought criticism of lack of patriotism from some quarters, but the deal which was being offered by Renault, was too tempting to turn down.

The 1960s was not the first time that the French manufacturer, Renault, had made an impact on the streets of London. Back in October 1906 the then recently set up General Cab Company had taken delivery of five hundred of Renault's two cylinder Type AG taxis, similar to those operating on the streets of Paris. These vehicles caused quite a stir and some controversy. Initially, the Renaults could only undertake private hire bookings, as they were not yet licensed for public hire. But their presence on the streets helped to advertise the company ahead of the cabs' official launch on 21 March 1907. The vehicles were to be fitted with meters and the drivers would receive a percentage of the takings, which the unions of the time opposed, due to the fact that other

An advertisement for the Renault Dauphine, conveniently
in a London scene. (Renault Ltd – Welbeck Archive)

operators were paying their drivers a set wage of five shillings plus two shillings in the pound commission. The unions argued that this new system would create a reduction in earnings for the taxi drivers.

Over fifty years later, Renault were back in the press under similar circumstances and with a keen eye on promoting their chic, four-door Dauphine saloon by means of a deal with Welbeck Motors, which would raise the profile of this French car. The first Dauphine had rolled off the production line in December 1955 at Renault's ultra-modern, Flins located factory in France. An earlier notable occasion in the history of the car's publicity was when Queen Elizabeth II accepted the gift of a British Acton-assembled Dauphine, when she visited the Flins plant in 1957. The car was painted powder blue and came complete with a set of chrome wire wheels and it was kept at the Royal Mews for many years.

The car's popularity was given a boost, when it won the 1958 Monte Carlo Rally outright and a Gordini-Dauphine version was driven by famous racing driver, Stirling Moss, in the same year from Marble Arch in London to the Arc de Triomphe in Paris (via Silver City Airways), as part of the 50th Anniversary Celebrations of Bleriots' first aerial crossing of the channel.

However, in reality, although standard production cars were known for their smooth, shock-absorbent ride they could be slightly difficult to handle, when cornering at higher speeds, due to the rear engine layout and the rear wheels being located by swinging half axles (but with no stabilizer rods) to provide independent rear suspension. Power was provided by an 845cc, 30 brake horse power (bhp) four cylinder engine, mated to a three speed gearbox and drum brakes all round.

Amédée Gordini (1899-1979), the French engineer of Italian descent, who had made a name for himself in the post World War II years as a David to Ferrari and Maserati's Goliath in single-seater formula racing, had joined Renault in 1956 as a consultant. His first commission was to produce a sporting version of the Dauphine, which bore his name and which was launched for the 1957 model year at the Paris Motor Show in October 1956. Within strict cost parameters, this version, the Gordini-Dauphine, was notably improved engine-and gearbox-wise. The same capacity engine offered 37.5 bhp by means of a modified cylinder head with twin water take-off and inclined valves, special inlet and exhaust manifolds with a larger carburettor and a sharper camshaft. Gordini managed to insert four forward speeds into the standard three-speed gearbox casing, which improved the car's tractability and on-road performance considerably on all Dauphines, including the new Gordini. Plate wheels replaced the hallmark 'spider' wheels, which the first version of the Dauphine had inherited from the Renault 4CV.

For the 1960 model year, the Dauphine-Gordini inherited the 40 bhp 'Ventoux' 845 cc engine, which had been updated by the factory with larger valves, improved camshaft, less restrictive manifolding, etc, for fitment to its new 'Floride' convertible. 'Aerostable' suspension (a system patented by J A Grégorie) was fitted to all Dauphines and Gordinis of this period for the 1961 model year. There were also short lived 'deluxified' Ondine versions of these two models (only the Ondine-Gordini was sold into the UK market, called the 'Gordini de Luxe'), which ran on slotted, plate wheels and offered four-speed gearboxes and plush interiors with reclining seats. By 1963 the Dauphine had sold over two million examples and it was eventually withdrawn from the market in 1965 after it and the Gordini had inherited the improved, four-speed gearbox and disc brakes from their younger sister, the Renault R8. The sole Gordini survivor was phased out in 1968, by which time its younger sister, the Renault 16 (launched in 1965) was revolutionising the mid-size saloon market, with its 'then new' hatchback.

The year of 1960 had been a difficult one for Renault, pressured by a depressed American market where it was reported that vast stocks of unsold Dauphines had built up. The company had been forced to lay off over one thousand employees at the time and it is possible that this could have played a part in Renault's decision to offer Welbeck Motors such a competitive deal on a large fleet purchase. Renault had offered a superior bulk purchase deal for the Welbeck Motors operation at £465 per vehicle (a standard Dauphine in March 1960 normally retailed at £716 new) and this included the 12-volt conversion kit, which was necessary to operate the radio equipment. The sporty, two-door, rear-engined Renault Floride was a popular seller at the time, so the company really wanted to be seen on the streets of London with its four-door Dauphine, thus demonstrating its completely different attitude towards sales and publicity in comparison to its British industry counterparts.

On 10 March, it was announced to the press that the order for the first two hundred vehicles had been placed and that this would eventually total eight hundred units by the end of the year, at a cost of £560,000. Renault's then UK chief, Mr Jean Ordner, commented to the press 'We are happy to have received this order, which is the biggest ever placed for a non-British car by a single British purchaser'. It was arranged that the first cars would be collected from the first week in April at the rate of five a day, bearing firmly in mind that the first two hundred had to be ready for the June launch of the 'pilot fleet'. Michael was authorised to place an order for a further three hundred cars afterwards. The General Guarantee Corporation confirmed on 4 April that they would be loaning the company a further £40,000 in four weekly instalments of £10,000 as the start up costs of the operation built up.

Incidentally, on the back of the Welbeck Motors order, another London-based car hire company, Roy Galway ordered two hundred Dauphines worth £140,000 for Whitsun delivery, citing a preference for the Renaults, due to their economical operation and low maintenance and fuel consumption costs. The publicity gained must have certainly made the deals worthwhile for Renault.

In reality, the Welbeck Motors vehicles would be red in colour and not the bright yellow which Michael had first envisaged. Conditions of the collection arrangement required Welbeck Motors themselves to collect the new cars with their own car transporter from Renault's premises on Western Avenue in Acton, West London for delivery to its own workshops in Taunton Place, which would keep Renault's costs down. The electrical conversion would be undertaken by Joseph Lucas Ltd at a cost of £5 10s per car, with the power system being upgraded from its standard six volts to twelve volts, thus permitting the use of the Pye Telecommunications latest 'Ranger', part transistorised mobile radios. The Dauphines benefitted from the fitment of the four-speed gearbox, as used in the higher specification Renault Ondine and Gordini models, in place of the earlier three speed unit, as this would better suit the cars in their role as minicabs. At the same time the important Halda Mk 6 mechanical taxi meters were installed. Usually costing £35 each, the units were discounted to £31 10s, and fitted by the suppliers for

Mk 6 Mechanical

HALDA TAXIMETERS

Supplied to Londons Horse Cabs 1908

NOW Supplied to London Mini-Cabs 1961

Mk 8 Electric

Halda Taximeters are fitted to almost every known make of motor vehicle— Rolls-Royce ...to motor scooter

HALDA

HALDA LIMITED Dept. B, 2 Brandon Road, York Way London, N.7 Telephone NORth 1207

Advertisement for Halda Taximeters. ('Car Hire' magazine May/June 1961 – Welbeck Archive)

an additional £2 per vehicle. Halda's magazine advertising of the time described its equipment as 'Supplied to London's horse cabs in 1908 – NOW – supplied to London's Minicabs in 1961'

At the start of operations, new Renaults would arrive regularly by transporter in order to be prepared for service. They were delivered in batches throughout 1961. With the orders being staggered as follows: 7, 11 and 19 April – *90 cars*; 9, 11, 15, 19 and 25 May – *89 cars*; 26 June – 22 cars; 11 August – *25 cars*; 19 September – *24 cars* and, what records appear to show as the last order, 25 October – *25 cars* thus making a total of 275 vehicles. No records have been found to confirm whether any further cars were ordered after this date.

The first vehicle ahead of the launch was registered 868 BLX, a Greater London County Council-issued number plate, which was used on many of the publicity photo shoots with drivers, Bill Buck and Jane Simpson, as well as general shots for other uses. The rest of the soon-to be-arriving cars were registered in batches of consecutive number plates, again issued by the Greater London County Council. According to some Welbeck Motors files, the numbers were 101 BYN through to 160 BYN, then with more numbers from 170 BYN to 191 BYN. The next batch of issues appears to start at 200 BYN through to 260 BYN, with further numbers showing in the paperwork as 272 BYN through to 294 BYN. The last batch is shown from 302 BYN up to 349 BYN, but with 351 BYN also being registered to a Renault minicab. It is likely that Welbeck actually took up all three hundred allocations from 100 BYN to the high 300s, which would tie up with their initial plans for the size of the minicab fleet. Unfortunately, the DVLA records for this area were destroyed a few years later, so it has been impossible to determine the exact numbers finally registered. Latterly, it appears likely that some of the Renault Dauphines from Welbeck Motors hire fleet were put into service in the later stages of the minicab operations, as Renault Dauphines from what was the separate side of the rental business have been listed as being used as minicabs: 749 ALU, 704 ALP and 771 ALU are examples, which were definitely recorded.

A brand new Renault Dauphine is seen at work.
(Graham Walker – Welbeck Archive)

One incident, which was not widely reported in the early weeks of operation, was of the Welbeck Motors' car transporter, which was fully loaded with new Dauphines and which became involved in what appeared to be a deliberate accident, with a London taxi cab suddenly stopping and causing the transporter to take avoiding action, which resulted in wrecked new cars and the transporter itself.

Although it had initially been expected that the Renaults would be used for a year and then replaced, in reality, some were said to be

A passing minicab is captured in traffic in front of an Austin
FX3 taxi. (Graham Walker – Welbeck Archive)

used for only six months, as some drivers absolutely destroyed them. It has to be said that, although the Dauphine was a very popular car with broad appeal in many markets, it was not really suited to minicabbing in London's busy streets. Clutches would fail and staff commented that they 'were made of paper' compared to a 'real' London taxi cab. Once a vehicle's use as a minicab was finished, its appeal was limited, because, when they were registered new, many of the number plates followed the reserved allocation of the BYN series.

Further, as they were painted red in colour, potential purchasers quickly realised that a late, second -hand, red Dauphine for sale in the south east of England could easily be a former Welbeck Motors' minicab. By having the cars repainted blue or white after being taken out of service, it was then possible to obtain in the region of £300-£400 for them on re-sale. Early costings by accountants indicated that the cars would be assessed as being worth £365 on disposal. Hugh Ruding Bryan, being in sales at Welbeck Motors at the time, had been given the task of disposing of the minicabs when their service life was at an end.

Indeed Renault had given Welbeck Motors an extremely good deal for this unusually large order of cars for May/June delivery. Looking ahead, as part of Welbeck Motors costings, estimates suggested that the value of a one-year-old, privately-owned Dauphine in the following June of 1962 would be £500; the value of a former self-drive car would be £450; the current March 1961 value of a one-year-old Dauphine, according to Glass's Guide, was then £465, which was the figure which Welbeck Motors would be paying for its new vehicles. A sales advert from 'The Motor' magazine in July 1962 lists Welbeck offering two 1960 models for sale. The first, a red model, had a sun roof and, being 'A car we are proud to offer', it was priced at £395. The other Dauphine, as the ad said 'Green, again in splendid condition and sold with three months' guarantee' was priced at £380. One wonders if the 'green' vehicle was, in fact, a re-furbished minicab, or an example from the separate car hire fleet. At the time, minicab examples, which were being sold off to the trade, were fetching in the region of £200 to £235 per car.

Quite a number of the Renaults passed on to Ray Thacker at the Colindale Service Station in Colindale. He was giving around £150 per car, whatever their condition and it was Ray, who eventually said, after taking dozens of them, that he could not take any more of them as the public had become wise to these prominent, red-coloured cars for re-sale. Welbeck Motors then started painting one or two, but Ray asked them not to paint any more cars, as he would repaint them himself, in-house at half the price. As salesman Hugh put it 'They were done with a catapult at twenty-five yards!'; he saw one at Colindale, where they had just masked the windows with paper and the paint sprayer had blown paint all over the car, with overspray on the wheels, too! And, if one opened the bonnet or doors, the car's original red shade was still there for all and sundry to see.

Chapter 10

Spin offs

Sometimes, a business creates a brand awareness, which is not always foreseen by the company itself. British haulier, Eddie Stobart is a good example of this and their smartly liveried lorries have become instantaneously recognised anywhere in the country in recent years. Perhaps, in a small way, Welbeck Motors garnered such an unexpected legacy.

The publicity which was generated from the launch of such a substantial fleet of new minicabs, brought additional publicity from areas, which perhaps, one might not consider. The largest tangible legacy of the Welbeck Motors operation is undoubtedly the Dinky Toys' own creation of the Renault Dauphine Minicab.

Publicity material printed in magazines and papers of the time for the new model stated 'For hire in London . . . minicabs, now part of the London scene, are standard production models of the famous Renault Dauphine. Advertising space is rented to manufacturers by the owners of these small taxis, and the Dinky Toys' model is a realistic miniature with six colourful posters – and windows, of course' The model, number 268, with a length of three and eleven sixteenths of an inch, was priced at 3/3d and was only available in the UK. The advertising stickers on the model differed somewhat from the real, in-service minicabs by having alternative advertising of popular brands of the time such as HP Baked Beans, Meccano on its nearside and Kenwood on the offside. But it did have 'Britax' Safety Belts across the rear engine cover, as on some of the real Minicabs, as well as Welbeck Motors' own branding on the side roof lines.

The original model of this Renault Dauphine as a saloon car had been launched in June 1957 for the French brand of Dinky Toys under a special agreement with Renault. In 1962 the mould for this model was now sent to Dinky's Binns Road factory in Liverpool in the UK to be created as the 'Renault Dauphine Minicab' and was produced over a four year period. These models are now sought after by collectors across the world, with mint condition examples commanding a respectable price.

On National Television, the venture continued to be of interest on shows such as 'Town & County', 'This Week', 'Tonight' and 'Town & Around'. The start of a new BBC TV series of 'It's a Square World', starring Michael Bentine, was screened on 28 September 1961 and in which he dealt with topics of the moment with a twist of his own brand of humour. This show's highlights included matters such as Russian space flight, what makes a Briton and Maxicabs – the taxi drivers' reply to minicabs.

Michael (Gotla, not Bentine) was invited to an event called 'Men of the Year' along with a wide assortment of other men whose achievements in 1961 had been reckoned to be worthy

NEW

For hire . . . in London

Minicabs, now part of the London scene, are standard production models of the famous Renault Dauphine. Advertising space is rented to manufacturers by the owners of these small taxis, and the Dinky Toys model is a realistic miniature with six colourful posters—and windows, of course.

DINKY TOYS No. 268 RENAULT DAUPHINE MINICAB
Length 3 11/16 in.   U.K. Price 3/3

DINKY TOYS

MADE BY MECCANO · LIMITED

AVAILABLE IN U.K. ONLY

Published by MECCANO LTD., Binns Road, Liverpool 13, England   Printed by John Waddington Ltd., Leeds & London

Dinky Toys' advertisement for the 'Renault Dauphine Minicab' (Welbeck Archive)

and newsworthy. It was held on 10 November at the Savoy Hotel in London and special guest and international Man of the Year, Mr Bob Hope attended the lunch, which was held to raise funds for the British Council for the Rehabilitation of the Disabled. The press made light of the fact that the Transport Minister, Mr Ernest Marples, was also being invited; perhaps he should have been seated between Michael and another invited guest, Mr Jim Francis, secretary of the Taxi Drivers Union, in order to be able to keep the peace. The Minister was apparently in such ebullient humour that he said, 'I wanted to sit next to Mr Michael Gotla, the minicab owner, because one of my constituents was knocked down by a minicab. I asked him 'Did you get the number?' He said, 'No, but I'd know the laugh anywhere!' The humour of the event continued in the Minister's speech, when he said 'People want you to get to the top in politics, but they don't want you to stay there'. Bob Hope told him consolingly, 'One day you are drinking the wine and the next day you are picking the grapes!' On the list of names which had been chosen for the Man of the Year accolade were: Richard Dimbleby, Robin Day, Dr Gerald Slot, the cast of 'Beyond the Fringe', Hardy Amies, RAF Corporal David Redman, holder of the Lorry Driver of the Year award, Sir Bernard Lovell of Radio Telescope fame, Mr John Powell, Chairman of Dome Press, Gordon Pirie, the athlete, and Michael Gotla of Welbeck Minicabs.

Clubs, societies and all sorts of organisations were interested in the minicabs project and it was known that Welbeck Motors were happy to share information about how their operation worked. For example on 11 November the London Appreciation Society, which arranged tours and talks of interest about the capital, visited Taunton Place to be taken on a tour of the operating rooms and behind the scenes, in order to see how the world of minicabs worked. Generally the society was in favour of the advent of the minicab, as comments on the advance programme had been made on how this new type of transport was one of the most exciting things, which had happened in London in recent years, and that 'All Londoners know the frustration of hailing cabs, which are not there or which are not already engaged, and the fruitless hours spent telephoning taxi ranks.'

One wonders if the scriptwriters for Pinewood Studios also saw something in the real life business played out on the streets of London when, in June 1963, the film 'Carry on Cabby' was released in cinemas across the country. A Peter Rogers production, it was filmed with a budget of £150,000 between 25 March and 7 May that year, starring many actors who had become household names such as Sid James, Hattie Jacques, Liz Frazer, Esma Cannon, Amanda Barrie, Kenneth Connor and Charles Hawtrey. The story was based around two rival 'cab' firms, one of which had set up using female drivers. Escapades, such as the two-way radios on the cars being tampered with, the female drivers tempting passengers to travel with them and other 'Carry On' innuendos, went without saying. And the Ford Motor Company wasted no time in seeing the value of promoting their new Consul Cortina Super on the silver screen, where it was billed as 'The small car with the big difference' on the promotion material for the film, in which this vehicle had been chosen to star.

Publicity poster for the film 'Carry on Cabby' in 1963. (Media Storehouse)

Chapter 11

Minicab fever across Britain

In other parts of the country, other minicab operators were starting up their own businesses, in many cases inspired by the Welbeck Motors' operation and its publicity.

The Minicab fever was certainly not confined to Britains' capital city. The 'Evening Times' in Glasgow reported on 6 March 1961 that the first minicabs should be rolling along their streets in June. Tom Sylvester, who already ran a car hire business in London, was planning for twenty-five Fiat Multiplas to run in Glasgow and he had ideas for Birmingham, too.

In early April, Clifford Taxis in Southampton announced that it would be taking delivery of twenty-five minicabs and that it would be recruiting twenty-five women drivers as well, who could be expected to earn between £9 and £15 per week. This effort was being made in response to reports that London-based Tom Sylvester

Mini taxis around the country! This example, with a
distinct two-tone livery, was operated by Mr J H Spence in
Edinburgh. ('Car Hire' magazine – Welbeck Archive)

would be basing twenty-five Fiats in this important South East town
from June. In the same month, in Edinburgh, Mr James Spence, a
taxi-cab proprietor was said to be replacing his taxi fleet with twelve
minicabs and that he hoped to be able to cut his fares by 50%.

On 1 September, Coventry gained its first minicabs. Five Renault
Dauphines were put onto the road by Mr Tom Lewis of Mini Taxis
to compete with the eighty existing taxis, but there were problems
with landline telephones not being installed, due to a maintenance
go-slow by Post Office engineers, although there was an assurance
from the chairman of the Coventry Taxi Proprietors Association
that there would be 'no trouble' with the new venture. However, it
did not take many days for someone to deliberately jam the firm's

Coventry's new Renault Dauphine minicabs, 'Mini Taxis',
seen on the first day of operation on 1 September 1961.
(Coventry Evening Telegraph – Welbeck Archive)

telephone lines late on a Saturday night, about which the firm could do nothing. The Coventry Telephone Exchange staff was praised by Mr Lewis for their help in assisting with the situation.

At the end of September, Manchester gained a fleet of thirty-six new Austin Seven De Luxe models (commonly known as 'Minis') operated by A. Charlesworth Ltd. The radio equipment for these vehicles was hidden in the boot and a roof rack was fitted for luggage.

There was similar news in Plymouth, where six new Renault Dauphines took to the road, operated by a Mr Michael White. Previously he had been a pig farmer and he had switched to amusement machines, before suddenly deciding to buy minicabs, having seen four identically coloured examples for sale at a garage and he had bought them on the spur of the moment. He had been inspired by the Welbeck Motors events and was, like them, charging one shilling per mile. Plans for more vehicles were already being considered.

Plymouth's fleet of six brilliantly coloured new Renault Dauphine minicabs, which have started working in the City – one of the first places in Devon and Cornwall to have them. (Western Morning News – Welbeck Archive)

Other London minicab operators were taking to the Dauphine, too. Carline of Wimbledon had decided in mid-July 1961 to begin replacing its familiar, blue and grey two-door Ford Anglias with the little French cars, placing an order, estimated to be worth £70,000. The Company found the smaller two-door Fords to be somewhat awkward for passengers and so they decided to switch over to the French vehicles, due to Renault's more competitive offer for a bulk order. Unlike Welbeck Motors, its cars would remain in their own livery of blue and grey and the company was quoted as saying 'they would not be splattered with money raising advertisements' and Carline hoped to have one hundred on the streets in between six weeks' to two months' time. In fact, when the last of the Fords had been replaced by late August, there were said to be fifty Dauphines operating. Carline also boasted that it had become Edinburgh's second minicab firm to launch, when, in September it put six Renault Dauphines onto the streets, thus emulating an operation in Bristol.

Furthermore, in Bath, a Mr John Smith of Minicabs (Bath) Ltd spoke of his initial success with a fleet of twelve Renault Dauphine minicabs which had enabled him to double the operational fleet in five months. He noted that his passengers came from a broad spectrum and included those travelling to work or even to school, peers, charwomen and dustmen. He reported a couple of incidents with traditional taxicabs but, overall life had been much less troublesome than on the streets of London.

In Kettering, minicabs appeared at the start of October, and the only remarks in the press was that they were in great demand and that local taxi drivers saying said that their arrival on the scene had made little difference at all to their trade. The owner of the minicab firm, Mr Starmer, did not foresee any trouble between minicabs and taxis in Kettering, 'That sort of thing doesn't happen here' he said. In fairness to Mr Starmer, he had been in the taxi trade all his life and it was his father, who had started one of Kettering's first taxi services in 1928! In Manchester, Mr B. D. Charlesworth put twenty-four bright red minicabs onto the road at the same time and he reported that he was 'snowed under with enquiries'. However, a spokesperson for the Manchester & Salford Taxi Owners' Ltd warned that 'We'll be keeping a close watch on them'.

In Newton Abbot, taking inspiration from the Welbeck Motors' story, a company called R.E.G. Motors started a new service called

Mr Harold Baldwin, Director of Carline Minicabs pictured
with one of his Renault Dauphines. ('Evening Dispatch',
Edinburgh 16 September 1961 – Welbeck Archive)

'Dinkikab' with two new Renault Dauphines. Previously, one of the
partners had attempted to apply for a hackney licence from his local
Town Hall, but he had been told that 'They couldn't remember
when the last one was issued, it was so long ago' and it was reckoned
that there was not enough business to go around. The demand for
the new venture's services soon proved the Town Hall wrong.

This story was repeated all over the county, in different places with different feelings and attitudes towards these new operations, with many starting in business during the month of August. But the troubles with the opposition in London did not diminish as the year drew to a close – in November, Sylvester Car Hire reported that of his fifty green Fiat minicabs, there were normally around five laid up due to smashed windscreens, slashed tyres, crumpled bodywork and ripped seats. Mr Sylvester felt that things could probably get worse, as, during the winter months, the cab trade was not as busy as in the summer months. He added that, perhaps rightly, the taxi drivers might feel that his firm was taking their livelihoods from them and that conditions might well get even worse.

Minicab exploits were by no means confined to our British shores. For example there is a fascinating story of similar happenings in Scandinavia which appears in the Danish book "Renault i Danmark – Mennesker, biler og markedsfØring" by Erich Karsholt & Henrik Stenholt (Strandbergs Forlag, 2014 ISBN 87.7717.227.4).

Late in 1961, a group of commercial vehicle operators in Aalborg purchased ten Renault Dauphines and established a taxi company under the name of 'The Mini-Cab Company', which offered fares at between thirty and forty percent lower than the prices charged by the established taxicab trade and by small car companies. The idea for the Mini-Cab Company was taken from, amongst others, London, where the monopoly of the classic London 'black cab' had just been challenged by the Renault Dauphines [of Welbeck Motors]. In that city, this had led to fights between the drivers of the new small cars and of the established taxis.

Similarly, in the North Jutland town of Aalborg, the established taxi drivers fiercely criticised the smaller cars as being totally unsuitable as taxis. More than this, they did not want the new company to employ female drivers. Lund Petersen, business manager for [the traditional] Aalborg Taxa, stated that 'With regard to women drivers, even though they may have professional driving licences, we are of the opinion that it would be irresponsible to send a woman out into heavy traffic in a small car such as a mini-cab'.

Co-owner of The Mini-cab Company', Kissy Haugaard dismissed this criticism. She herself had had a professional licence for some years and she considered it insulting that irresponsibility should be linked in some way with female drivers. She stated that the minicabs were cheaper at the outset, which resulted in the overall costs being less than with the larger vehicles. A Renault Dauphine could be purchased for less than half the price of a Mercedes taxi.

The drivers of the established taxicab companies were not happy with these new developments. They went as far as going on strike and lobbying officials of the Aalborg City Council and of the Christiansborg City Council to prohibit the use of the smaller hire vehicles, which threatened their livelihood. According to the media, this played into the hands of the minicab drivers, who organized their own drivers' union. But the politicians did not intervene, and, instead, spread the idea of the minicab service to other cities in the country.

The Mini-Cab Company was a great success in Aalborg. In the following year, more Renault Dauphines were purchased so the company bought a total of twenty-eight of these French cars. There were no taximeters in the first cars, so the fares were agreed between the drivers and the customers. Kissy's son, Preben Haugaard recalls that a short trip cost three krøner, a medium trip cost five krøner and a longer trip cost seven krøner. A taximeter was fitted to the later vehicles. At one point taxi No.17 and taxi No.18 were both seriously damaged, one at the front and one in the rear. The local Renault dealer, Svend Moller, suggested that he should take the good ends from both vehicles and weld them together into one good vehicle, which he did. The 'new' car combined the numbers from the two damaged cars, so that it came to be called taxi No. 35 – a creative solution, which would not be tolerated nowadays!

Preben recalls that one of the blue Renault Dauphines topped 300,000 kilometres. The company later used Renault Eights, Renault Tens and Renault Sixteens as minicabs, but never in the same quantities as the Renault Dauphine.

In 2006, the Aalborg Mini-Cab Company became a subsidiary of Aalborg Taxa and, shortly before its fiftieth anniversary in 2011, it was completely absorbed into the parent company and the 'Mini-Cab' name disappeared.

Chapter 12

Trouble ahead

The crisis within the taxi trade would eventually come to a head. How the minicab operations were able to continue to operate when they were more loosely regulated than the taxi industry's heavily regulated system and in some cases minicabs were also seen plying for hire.

As the reader can deduce, the operation was not without its day-to-day problems. An unhappy taxi industry had arranged a meeting long before Carline and Welbeck Motors' operations were due to start and it took place on 6 February 1961 at Fulham Town Hall. Over two hundred taxi drivers were present and they had been invited to attend by the taxi drivers' branch of the Transport & General Workers Union. At the meeting, no decision was reached on a course of action to challenge this new, unregulated type of operation at this stage, although it was agreed that action must be considered at a higher level after the legal position had been made clear. Other meetings followed, specifically one on 7 March at Seymour Hall,

Marylebone, at which an estimated three thousand drivers attended a protest meeting, where they were kept informed of talks on the on-going issues between taxi-cab proprietors and union leaders. It was reported that nearly two thousand five hundred parked taxis created a monster traffic jam, while the two hour meeting took place. Police had to sort out the traffic chaos for well over half-a-mile around the hall, and, in the West End, hundreds of theatre-goers were unable to hire a taxi to take them home. Actually, Michael attended the meeting himself, in order to hear what was being said, and, in a subsequent interview with 'The Times' he stated that he did not propose to fight a war, that he had nothing against the normal taxi and that there was plenty of room for an additional service. Michael continued that there was a shortage of taxis throughout London and he suggested that there was enough existing business to justify the use of another one thousand vehicles.

It was obvious that minicab firms would be in for a difficult time when a 'Sunday Times' reporter published an article on 28 May revealing how the Taxi Drivers' Union had sent out a bulletin to its members, advising them how to harass London's minicabs. It suggested that, whenever a minicab was seen parked, a policeman should be called and an allegation made that the vehicle was plying for hire. Further, the document mentioned that the taxi drivers might have brought the competition on themselves and they should improve their courtesy and consideration towards passengers.

As if the troubles with opposition from the taxi trade at this early stage were not enough to contend with, Sir Wavell Wakefield, M.P for St Marylebone was campaigning in the House of Commons to stop the noise and the congestion, which was being caused by the minicabs around the Taunton Place base, which was now operational. In the House on 20 July 1961, the MP told Mr R A Butler, the Home Secretary, that minicabs, which were garaged around Taunton Place and at nearby Boston Place and Balcombe Street, were causing a disturbance to residents in those areas. Sir Wavell had asked Mr Butler what steps would be taken to stop 'The banging of doors at all hours of the night and revving of engines', but when the local newspaper 'The St Marylebone Record' decided to interview the local residents, the message was generally unanimous – 'Let them alone'. Many residents were indignant and refused to comment other than saying 'Too much is being made of the affair'. Residents opinions ranged from 'The company has created employment for two hundred men' to 'I have to earn a living and so do the minicab drivers, so they deserve an opportunity'.

Michael's solutions to these early problems ranged from stopping night shift changes at the garage, dropping the use of the loud speaker system to address drivers in the streets, as well as organising two-man driving teams, which meant that one minicab would be based at one or the other driver's private home. He wanted to co-operate with requests from residents, but he had also made plans to keep two hundred minicabs at Lord's Cricket Ground in St John's Wood at the close of that year's cricket season. Apparently, according to newspaper reports of the time, fifty cars had already been using the ground's car park, and it was hoped that the MCC would grant permission to build a radio control hut there.

A number of taxi drivers decided to boycott the 'Westbury Hotel' in New Bond Street, after taxi drivers had spotted minicabs pulling up on the forecourt, in order to collect their fares and it was felt that preference was being shown by the hotel to minicabs over the taxis, and so, for nearly an hour, the taxi drivers picketed the hotel and refused to pick up guests. The police arrived and two taxi drivers stated that they were members of the London Cab Trade Crisis Committee and they had had an interview with the hotel management, at which it had been agreed that minicabs would not be called, unless specifically requested by their guests.

At around this time Michael made a separate announcement that the fares for the minicab venture might have to be raised in September, owing to a rise in the petrol tax. Michael stated that takings were rising, but that costs were continuing to increase, these being due to the continuing difficulty in attracting sufficient drivers. Reports indicated that there were three hundred cars on the road, but that not all of them were in constant use, due to a lack of drivers. In fact, an extra twenty-five minicabs were put onto the road during September, which did result in a slight increase in revenue.

Matters became more personal for the operation's financial backer, Mr Isaac Wolfson himself, with the on-going problem of the taxi opposition being thrown very much directly at him. Early in September, he was obliged to make Welbeck Motors aware that his three sisters, a Mrs Steinberg, a Mrs Williams and a Mrs Jay, all living in London, were being disturbed during the night by 'dud' orders for minicabs. They were being contacted by Welbeck Minicabs and asked whether they had ordered a minicab, which, of course, they had not. The memorandum from Mr Wolfson's office was tactful and said that the person responsible for the booking would have been either a taxi driver or a disgruntled ex-employee of Isaac Wolfson, and that such actions were giving the ladies cause for concern. Their addresses were passed out to the staff in the control room in order that these calls could be ignored, because they were a practical joke.

An internal memorandum was sent out to all staff at Welbeck Motors on 28 September 1961 from the office of the Managing Director stating that 'From this week, I regret that your wages have been cut by 10%'. The change was universal and applied to all salaried office staff and to staff earning over a certain figure in all the divisions of the business. The memorandum continued by revealing that many other economies were being made, including a reduction in staff and premises and that the present recession in the motor trade had hit the company very badly, particularly the fall in the value of the hire fleet. It went on to say that the Managing Director hoped that the situation would be under control by January of the following year (1962), when the company could move forward once again. Michael explained that these difficulties were being experienced by all companies in the motor trade, from the largest to the smallest and, like everyone else, the directors would have to run the business as efficiently and as economically as possible during what was likely to be a rather miserable winter. He ended the memo by stating that 'Whilst companies such as ours are the first to be hit by this recession, please bear in mind that we are also generally the first to get cracking again, at the slightest improvement in general motor-trade conditions.'

Michael had been forced to sell off what he described as a 'considerable portion' of the self-drive hire fleet and to close two of the car hire branches, which, as he explained to the board, were the largest part of the losses shown in the accounts, although this action would also produce a significant reduction in expenditure over the next few months. Michael also pointed out to the current board that the company was in serious need of immediate liquid capital.

Michael appeared on television again in the middle of October, and he was later congratulated by Mr Finnie of the Corporation in a board meeting on his 'extremely successful television appearance' which 'could do nothing but good.' But Mr Finnie asked that as much restraint as was possible should be exercised in all interviews with, and statements to the newspapers. He stressed the importance of 'playing down' the connection between Welbeck Motors and the General Guarantee Corporation Ltd, and even more so, that connection between Mr Wolfson and the minicabs.

Mr Finnie appreciated that, in the eyes of the press, the Corporation and Mr Wolfson were virtually one and the same thing, and that it was impossible to prevent the papers linking the two with minicabs but he asked that all possible restraint should be used in giving information to the papers.

The difficulties with opposition from taxi drivers flared up once again in a big way in early October, when more than two hundred taxis jammed an area around Leicester Square, Coventry Street and Piccadilly Circus, in an attempt to prevent minicabs from approaching the 'Prince of Wales' theatre, where Sammy Davis Junior was giving a late night charity show. Twenty minicabs had been hired in advance to pick up staff from the theatre at 03.30 am, but taxi drivers claimed that another twenty minicabs were also calling at the theatre, in order to pick up fares, who had not ordered in advance. Scores of taxis circled Leicester Square constantly, thus penning in the minicabs, which were parked by the central pavement. A Mr Bill Hooper, stage door keeper at the theatre, reported that the situation started about forty-five minutes before the end of the show. He stated, 'You can't blame our staff for ordering minicabs; they wanted to pay as little as possible'. The police arrived and ordered the taxis to keep moving, so that the minicabs were able to pick up the staff and they also checked that the minicabs had been booked in advance.

Situations beggaring belief were witnessed by the public on numerous occasions. One customer, who had pre-booked a Welbeck Minicab at midnight to take him from his home to a hotel, came outside to find a crowd of people trying to push the minicab over. To make matters worse, the person, who had hired it, was accused of trying to do taxi drivers out of a living by hiring the minicab!

In a letter to 'The Sunday Telegraph' on 16 November, Michael had written to a Mr Bennett, following their telephone conversation in response to a factual report which had been published by that paper on 22 October, covering the progress of Welbeck Motors' operation with special reference to the troubles it had been suffering, as a result of hooliganism. What had incensed Michael was that, once again, as he put it 'A newspaper nowadays has only to mention minicabs, and as sure as summer follows spring, it receives a letter from Mr Francis, Secretary of the London Motor Cab Trade Joint Committee'. Michael questioned Mr Francis' latest comments and he had decided to begin listing the most recent court cases of offences by taxi drivers against Welbeck Motors' staff. Michael said that he had been reluctant to discuss the taxi war, as he felt it was bad publicity for minicabs and that it had had little overall effect. He did agree with Mr Francis' words that 99% of the London taxi drivers were blameless, but he felt that it was strange that it was always Mr Francis himself, who would invite him to give documentary evidence of the violence. Michael made reference to a television interview, where he had been obliged to do so, because Mr Francis had already stated that the violence had been invented by him for publicity purposes. At the end of Michael's letter, he concluded by saying that there might be every sign that this phase of violence was dying down, hopefully because the responsible leaders in the taxi trade had regained control of the tiny minority.

It was a huge surprise to the public, when a story broke on the front page of the 'Evening Standard' on 24 November 1961, which was headed 'Chief Quits in Mini-Cab Dispute'. The story revealed that Michael Gotla was giving up his office and resigning from the board. It went on to say that 'R.S. (Jock) Walker has acquired all the shares in Welbeck Motors previously held by The General Guarantee Corporation Ltd and he would in future carry out the joint function of chairman and managing director.'

Mr Isaac Wolfson resigned as the company's chairman, and a Mr E M Cumming-Fuller was co-opted to the board. At this time Mr Walker already owned 39% of the share capital, and Mr Cumming-Fuller, a chartered accountant of the corporation, owned 10%.

An internal Welbeck Motors memo in the September of 1961 had already shown that the Wolfson family was reacting to stop its association with Welbeck Motors and later internal correspondence suggested that the General Guarantee Corporation felt that the trading situation was getting out-of-hand under Michael's direction. One key moment which may have brought matters to a head, was when Michael gave an interview to Fyffe Robertson of the 'Sunday Times' in October. It appears, from surviving papers, that Michael had been upset by what he considered to be an unwarranted curtailment of his duties to publicise the minicab project, and, he went on to inform his interviewer that a fleet of three thousand minicabs was being considered and that the finance for this project would be forthcoming from Isaac Wolfson. An article also appeared

Renault Dauphine minicab in traffic on A23 at Bolney,
West Sussex, 5 November 1961. (John Skinner)

in 'The People' publicising the seamier side of 'Operation Minicab',
for which Michael had issued a writ for libel against the paper on 26
October and this culminated in the withdrawal of the Corporation's
support, and Michael's resignation.

Before the board meeting on the afternoon in question, Michael
had been quoted as saying that 'This is a basic disagreement in the
boardroom about the policy of Welbeck Motors, and it could easily
blow over or not. We will know tonight. Obviously there have been
big differences in the Welbeck Motors board room, and it should
be sorted out there this afternoon. The future of our minicabs will
not be affected'. But, behind the scenes, Mr Walker was called upon
to take over the management and, furthermore, the Corporation

required Mr Walker either to purchase its 51% shareholding for a nominal consideration or to give them in return the right to purchase all the shares in Welbeck Motors for one penny per share, in the event of a receiver being appointed. Mr Cumming-Fuller would hold 1,500 shares and Mr Walker 6,150 shares plus the 3,675 shares, which he had bought from Michael. The two shareholders could not sell, assign or charge, without the Corporation's consent, any of the shares, whilst the company was under the charge of a debenture. Unless Mr Walker was prepared to act accordingly and to bring the situation under control, the Corporation threatened to liquidate the company under its power of debenture. An agreement was drawn up in November between Mr R S Walker and the General Guarantee Corporation, which was created by the latter, because they felt that if Mr Walker failed in his task, the Corporation would wish to regain control of the company. The statement to the press did not reveal the true position that, in fact, Isaac Wolfson's Corporation were still behind Welbeck Motors. Mr Walker also had to agree to the following statement for immediate release to the press:

'Welbeck Motors Ltd announces that their Chairman has acquired all the shares in Welbeck Motors Ltd held by the General Guarantee Corporation Ltd. The Corporation's representatives have resigned their directorships. Mr M F Gotla has relinquished the office of Managing Director and has also resigned from the board. Mr Walker will, in future, carry out the joint function of Chairman and Managing Director and Mr E M Cumming-Fuller M.A., F.C.A., has been co-opted to the board'.

The press, not knowing the true state of affairs, printed what it thought were to be the facts. For example, an article appeared in the 'Cardiff Western Mail' on the following day, with the headline

'Mr Wolfson quits Minicabs', in which it stated that Mr Isaac Wolfson had sold his interest in the London firm and that Michael Gotla had resigned as Managing Director and from the board of Welbeck Motors. It went on to say that, before the announcement, Michael had said that 'the credit squeeze had hit him badly' and that 'some of my backers are getting a little tight with money'. The announcement roused much interest in the City, as, earlier in the week, Mr Wolfson had been involved in a £2,250,000 deal with a Mr Max Joseph and 'Mount Royal Hotel' shares, which gave rise to speculation that he might be realising some assets outside his normal line of investment, with a view to making further take-over bids in stores and trading.

As has previously been stated that Mr Wolfson was embarrassed by the minicab 'war' and that, with what was thought to be over 60% of the London taxi men being Jewish, he was being given a hard time over his backing of the Welbeck Motors operation. The General Guarantee Corporation, like many other large-scale finance companies, operated in a much less public way, and the constant (and mainly) negative publicity through the difficulties with opposition from the taxi trade was harming his reputation.

A statement appeared in 'The Times' newspaper on 25 November 1961, in which Michael was quoted as saying, 'I have been grossly misrepresented and as a result, the hate of the whole taxi trade has been levelled against me. If cab drivers think my resignation is a victory, then they are mad'.

On 29 November, a mass meeting by a 'Crisis Committee' was held at Speakers' Corner, Hyde Park, which was followed by a lobbying of MPs calling for a forty-eight hour strike by taxi drivers on the same day. This action was taken by an unofficial direct action committee, to which the Unions were opposed. One thousand drivers were said to have attended this open air protest meeting. Chairman Mr Harry Slowman said that the majority of attendees were in favour of the action, which arose following the sentences of three months' in prison being imposed on two taxi drivers, with disqualifications from driving for three years on charges of dangerous driving, in connection with an incident involving a minicab.

The situation at the Welbeck Motors' business remained difficult. As the festive period drew nearer, a letter was sent to all staff warning them that, unlike previous years, when staff bonuses had been issued to coincide with the Christmas period, that year the company would not, unfortunately, be in a position to repeat the gesture. Apart from the extreme difficulties of the credit squeeze, the heavy depreciation borne by both hire fleets and the cost of mounting 'operation minicab' meant that severe economies had to be made, with setbacks in income for the staff, with numbers thinned and with the organisation and its overheads severely pruned. However, the letter did finish on a positive note stating that every effort was being made to put the company once again in a fighting position and, with teamwork, it was hoped to move forward to a more settled phase in the company's affairs.

Eventually Air France, one of the main advertisers on the vehicles, decided to pull out of the venture due to the negative publicity and this started a downward spiral of what was to become thousands of pounds of lost advertising revenue, which was fundamental to the viability of the operation.

Even though Michael was no longer connected with Welbeck Motors, he took the time on 8 December to write a reply to a letter, which had been sent to him by Viscount Prestwood just over one month previously. In it, Michael apologised for being inexcusably discourteous and explained that, when the letter arrived he had been immersed in major board room differences, which had culminated in his resignation as Managing Director. He said that he was passing the correspondence on to Mr Walker, who, in due time, would be writing to him. The Viscount's letter itself contained what he had observed as being some of the faults in the minicab service, and, more interestingly, some detailed suggestions for overcoming them, in order to help to sort out the difficulties, which Welbeck Motors was still facing. It was a shame that when the letter arrived, it was at such a difficult period for the business, because Viscount Prestwood had put forward some truly interesting ideas and thoroughly prepared pieces of advice, which would certainly have given the board something to discuss and which might well have benefitted the operation.

In December 1961 Welbeck Motors was forced to put up its fares to 1s. 4d per mile (an increase of over 33%), as due to increases in fuel, insurance, road fund tax and higher wages, it was no longer possible to continue at the original trial fare of one shilling per mile. The increases were said to have no connection with Michael's departure from the business a month earlier. At this time, it was also decided to stop issuing free uniforms and to tighten up on the issue and use of petrol and the expenditure on coachwork rectification necessitated by high accident rates. Pye, the radio providers, had their night service package cancelled, which saved a further £200 per month. Further savings of another £200 per week were made in the control room and of some administration costs at £40 per week. The drivers were being closely vetted and, as a result, one hundred and fifty of them were discharged. Although this automatically reduced

This wintery scene at Taunton Place captures both accident damaged and out of service Renault Dauphines. (Stanley Roth collection)

the operating fleet to around one hundred and forty vehicles, it did enable the service department to begin to catch up with the backlog of some badly needed maintenance on the minicabs. Assisted by the increase in fares, the income of the reduced fleet kept pace with the previous November and, in fact, the January income was £30,402.

In the self-drive hire business, branches were closed as rapidly as the company could disentangle itself from existing arrangements. The Slough branch was kept open at the time, as it operated much more economically than any of the others, and it was run by trustworthy personnel. It was anticipated that by the following

Withdrawn Renault Dauphine minicabs, soon to be
disposed of. (Stanley Roth collection)

March, this policy would have saved some £8,000 a year in rent
alone, not to mention the cost of repairs, petrol, wages and all the
other expenses, which contributed to the operational overheads.
The wage bill was also reduced by some £5,000 per annum. The
car hire fleet, which had numbered some three hundred and fifty
vehicles during 1961, had dropped to two hundred by the following
February and it was planned to allow this to ease to one hundred
and fifty by the end of April.

The sales department's efforts were concentrated on reducing and
re-balancing the stock; on focusing on the export business, which
was an area where Welbeck Motors had previously performed very

A written off former Welbeck Motors Minicab awaits
an uncertain fate. (Stanley Roth collection)

well and on continuing in the second-hand car market, as the company's capital was already overcommitted in other areas. By offering a 5% discount arrangement on its contract hire vehicles, Welbeck Motors could recommence trading, but, due to the volume of business having to be restricted, the profit, at that time, was still insufficient to balance the outgoings, and to allow the maintenance of the fleet.

In the property portfolio, the company made substantial savings. A car park, which was being rented in Finchley for £700 per annum, was relinquished. The parking at Lords Cricket Ground, which was costing a whopping £5,000 a year, was cancelled and various mews properties were either sub-let or given up, which saved in the region of £900 to £1,000 per annum. At the Welbeck Motors building itself, Flat No.1, which was a company office, was let out for £485 per annum.

Michael's former Flat No.2 was assigned at £550 per annum and Mr Walker's own flat was assigned at £525 per annum. Such re-arrangements made for somewhat cramped accommodation at both Crawford Street and at Taunton Place.

Welbeck Motors had introduced a voucher book for customers, who would rather use this facility than use cash; it was called 'Minicab Money'. The book consisted of forty vouchers, twenty at a redeemable value of four shillings each, and the other twenty at one shilling each, with a total value of one book being £5. The inside of the book cover explained that 'for the convenience of customers, these vouchers can be used to pay the fare, and tip the driver too, if you feel like it'. The vouchers were valid for three months from the date of issue, and, if the person had not used them up before the end of the period, they could be returned to the accounts department

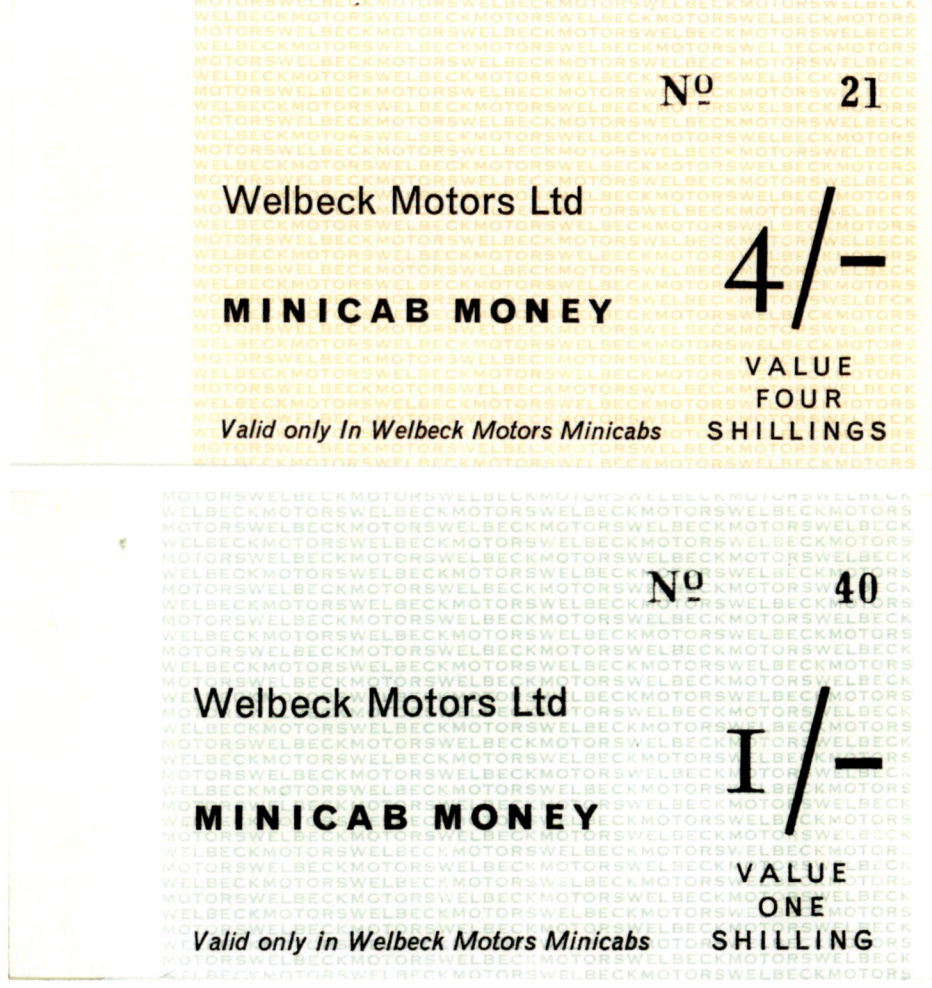

Minicab Money Vouchers. (Welbeck Archive)

at 109 Crawford Street either to be redeemed or for a new set to be issued.

On 26 February 1962, Welbeck Motors' solicitors, Amery-Parks & Co wrote to Mr Walker with a press cutting, which had been previously forwarded to them and which they had been considering;

it had first been published in 'The Times' on 17 February entitled 'Plying for Hire'. A case had been brought before the High Court of Justice involving a minicab owned by Radio Taxis, Liverpool, under section 45 of the Town Police Clauses Act 1847, in which the minicab had been parked outside a hotel in Liverpool on 15 April 1961, and, after fifteen minutes, a policeman approached the vehicle and questioned the driver. He replied that he was waiting for a call. In this case, the Lord Chief Justice had decided that he was satisfied that 'it was possible for a driver to be plying for hire when his car was standing in the street whether at a licenced stand or not'. His Lordship had continued by saying 'it was the essence of plying for hire that the vehicle in question should be on view, that the owner or driver should expressly, or by implication, invite the customer to use it, and a member of the public should be able to use that vehicle if they wanted to'.

Amery-Parks' letter said that this should be of some concern to Welbeck Motors, and that they had already picked up the report and were considering its relevance to their case. They were attempting to get a fuller report of the case in order to see how the facts might affect cases, which Welbeck Motors themselves had pending. Recent previous cases were noted and which showed the different attitudes of the magistrates in their different courts on hearing cases brought before them. Welbeck had numerous cases, but, in many of them, when an accusation of plying for hire was suggested, the minicabs had, in fact, been pre-booked via the Control and the drivers were legitimately collecting their customers without being hailed in the street.

On 1 March 1962, Mr Walker introduced a bonus scheme whereby a £15 per month bonus would be credited to a driver's wage card from that date if the employee had met the company's requirements, and the credits would be renewed in each four month period. The qualifications for payment (less deduction for income tax) were:

1. No accidents of any description, whether the driver was to blame or not.
2. No loss of equipment (unless paid for).
3. No motoring offences.
4. No complaints from the public.
5. No disciplinary action necessitated.
6. Still in the company's service.

The scheme made clear that, should the employee take care of their equipment, protect the company's good name and its interest in the vehicle they used, then they could expect to increase their earnings substantially. Their record card would indicate the quality of the entitlement, but, in only very exceptional circumstances, would an appeal be considered. At the same time, staff were given personal accident insurance for one year, and unless management heard to the contrary a sum of 15/-, which was their contribution towards the premium, would be deducted at the rate of 7/6 per week for the following two weeks. At this time, management was also taking stock of the cost of damage (willful or otherwise) of the cars and what had been happening to the drivers, since the operation began.

One hundred and six windscreens had been broken (at a cost of £6 each), Ninety-six tyres had been slashed (at a cost of £4 each), Thirteen cars had been damaged bodily (some deliberately overturned), five radios had been stolen, two radios had been maliciously damaged beyond repair and one car had been burnt out and written off. Two drivers had received injuries, which necessitated hospital treatment, and a further ten to twelve drivers had sustained minor injuries. Hoax phone calls put through to the control room in the early days was estimated at one in fifteen, and, in the early part of 1962, there were still between fifty and sixty hoaxes every day by unknown persons, advertisements for cheap unfurnished flats to let, items for sale, etc., for very reasonable prices, either in the press or in shop windows, using Welbeck Motors' telephone number. Bogus advertisements had been seen on newspaper boards in Belgravia, Chelsea, Pimlico, Paddington and Maida Vale.

In May, an interesting call, which was not a hoax, came through to Mr Walker from Mr Anthony Barclay of Jack Barclay Ltd, the renowned main London Rolls-Royce and Bentley distributors. It proposed that they should supply Welbeck Motors with Fiat Multipla taxis, if the opportunity arose for fleet replacement, but this offer was not pursued.

In April 1962, Isaac Wolfson called for a report on the Welbeck Motors' position. The report recognised the decisions, which had been made over the previous months by Mr Walker and it approved them. However, Mr Walker noted how 'touchy' these Corporation men were about any adverse publicity regarding Isaac Wolfson's connection with Welbeck Motors, and it was about this time that the first indications emerged that the Corporation felt that the business should be sold. In May, a Mr Simon Coorsh had been deputised by Isaac Wolfson to 'give Mr Walker a hand' at Welbeck Motors; At the time, he was the managing director of the General Guarantee Corporation, not just the Hire Purchase division thereof, but in its entirety, i.e. the Drage Group, the second 'Great Universal

Stores'. Mr Coorsh was welcomed to the board on 26 June and in the following month his plans emerged and they were different from those, which Mr Walker had envisaged.

Mr Walker had wished to keep the business trading, whereas Mr Coorsh did not. There was continued pressure to sell the business, although this was denied in certain correspondence.

In a report for the Corporation, prepared and dated 13 April 1962 by their accountants Finnie, Ross, Welch & Co, the teething troubles and substantial losses incurred by the minicab project were firmly blamed on the conflict with the taxi trade, by the dislocation of the service through numerous hoax telephone calls, by trades union action to bring pressure on the advertisers on the minicabs themselves, thus causing a loss of revenue fastened into the original budget, by increases in petrol, oil and tax introduced by the Chancellor, by staff problems and by the overwhelming public response to the service, with which the organisation simply could not cope, to the ultimate frustration of the public.

At the time the Corporation's accountants felt that there were three courses of action, which needed to be taken:

1. Full consideration of exploring all ways of channelling the Welbeck Motors losses into the General Guarantee Corporation, thus providing it with a considerable tax saving, although consideration would have to be given as to how this would affect the results of the hire purchase division.
2. The Minicab project should be closed and its fleet sold as a whole to fleet users of vehicles or, alternatively, over the next few months the minicab fleet should be quietly disposed of in the trade and the minicab operation closed.

3. Someone should be given the authority to sell Welbeck Motors
    Ltd and its car rentals subsidiary to another motor dealer in such
    a way that the loss to General Guarantee Corporation Ltd would
    be minimised.

The potential was still there for a viable minicab operation, as
the hire charges from the start of operation until 31 December had
shown revenue figures of £173,648 for the first 197 days with an
average of £880 a day. During the two first two months of 1962 this
daily average had risen to £990 with the minicab's income rising still
further, until just after this period, when the figures were around
£1,300 to £1,400 per day. At the time there were two hundred and
seventy five potentially operational cars with a recorded number of
just over the two hundred being mobile.

In the House of Commons on 16 May, a much discussed Private
Member's Bill was introduced by Sir Hugh Lucas-Tooth, which was
termed 'The Hackney Carriage Bill'. Its object was to amend the law
in relation to Hackney Carriages in London; to impose restrictions
on the use of radio communication for hiring vehicles and on the
description of vehicles for hire and for purposes connected with
such matters. Attention was brought to the fact that the 1869
Act, virtually unamended, was still law, and that it dated back
to the days of horse-drawn cabs and prior to the invention of the
internal combustion engine. Promoters of the Bill agreed that
the first matter should be to define what 'plying for hire' meant. The
expression in the 1869 Act was considered to be undefined and its
meaning extremely doubtful. Mr Rupert Speir MP in opposing the
Bill asked the Commons to believe that this was not the innocent,
harmless and helpful Bill as was being suggested. He went on to say

that 'It is high time the winds of change blew through the Public Carriage Office. I can imagine what it looks like, festooned with cobwebs, strewn with horse bags and with an all-pervading air of horse manure'. He went on to say that, although he criticised the taxi service, he was neither 'gunning' for all the London taxi drivers, nor that the present minicab drivers were incapable of improvement nor that their service could not be better. The debate coincided the same day with the regular ITV feature programme 'Dateline London', in which the ten minute programme dealt with a debate on the Bill. Both Sir Hugh Lucas-Tooth and Mr Walker appeared, and Mr Walker said that, because he wanted clarification of the law, he welcomed the Bill. He said that he was tired of ill-informed and baseless charges of plying for hire.

Welbeck Motors were in the national papers once again on 1 June, when an appeal was heard at the High Court of Justice in the case Rose v Welbeck Motors Ltd and another. This particular complaint had occurred on September 29, 1961, when taxi driver, Mr Emmanuel Rose had witnessed a Welbeck Motors Minicab parked in a bus lay-by. Mr Rose decided to speak to the driver of the minicab and, as a result of the conversation, the police were called. A bus arrived so the minicab pulled further down the road by about ten yards, and stopped. When the police arrived and questioned the driver, he told them that he was waiting for any job, which came up in the area via the radio. The case had been heard on the previous November 8, but, based on the facts, the Justices decided that there was no case to answer and they dismissed the case, but this decision was appealed by the other party. The sole question for the Court came down to whether the essential elements of the offence were present under section 7 of the Metropolitan Public Carriage

Act, 1869, and based on these facts, was there *prima facie* proof of those essential elements? Considerations as to whether the minicab could be seen by the public and as to whether the vehicle could be hired by the public. The vehicle's bright colour, the advertisements, and the fitment of radio equipment in what was, in essence, a car and the fact that the driver had been waiting for fifty minutes in a place where local buses would turn around with passengers boarding and alighting, and the fact that the driver left and then came back, did not help the situation.

The further ramifications for this High Court ruling were, with the essence of the cases that the minicabs were plying for hire being proven, that the advertisements, which were carried on the vehicles were deemed to constitute part of the visual awareness to the public. Lord Parker, the Lord Chief Justice, said that the advertisements on the minicabs meant they were plying for hire and thus breaking the law. The adverts he said meant 'I am a cab of Welbeck Motors. I am one of those minicabs and I AM for hire' The press quickly published articles about how a radio message was said to have been flashed to all the drivers of the minicabs stating 'Drive to a side street and take off your advertisements – quickly!' Within a couple of hours the huge task of stripping the Renaults had been completed and they were now operating in a plain red livery, looking exactly like a private car.

In the 'Daily Mail' newspaper (1 June), Commander William Milner, the joint General Manager of Welbeck Motors, was asked to comment and he said 'We don't mind really, though the ads did make the cars look distinctive. The colourful advertisements were taken off some time ago, the only ones remaining were those giving our name and phone number, but we are well enough known now to do without them'. He finished by saying that 'We did want these old fashioned laws about plying for hire cleared up and that's what's happening. It's all for the good'. Another minicab firm, Carline, tried what was described as 'a cute move' by displaying on one of its vehicles red and yellow diamonds and Union Jacks with the words 'this minicab is not for hire' in an attempt to draw attention to itself in an unconventional way. Even though the ruling meant that no lettering could now be displayed on a minicab, 'The Steering Wheel' magazine debated in its editorial pages of the 23 June issue whether the issue of the ruling was still insufficient and ended with the vital question 'Are minicabs, by virtue of their distinctive appearance (red in colour and exhibiting a radio aerial), plying for hire, whilst standing about in the streets?'

On 25 June 1962, Welbeck Motors was obliged to increase its fares once again, this time by a further 25% to 1s. 8d per mile. Based on a six mile journey, this increase now made a minicab more expensive to hire than a traditional taxi. Higher insurance rates, related to their early claims record, were due to be implemented, and these were blamed for the increase. In fact, the evidence of the Corporation's behind-the-scenes involvement in the on-going costs of the operation were evident, when the insurance brokers Price, Forbes & Co Ltd met with Simon Coorsh in Glasgow, at which meeting he took the liberty of discussing the renewal of the policy through Norwich Union. He proposed to insure a fleet of two hundred and fifty minicabs on the basis of third party, fire & theft cover only at a cost of £90 per car with a £100 excess. Mr Coorsh's plan was in accordance with the Corporation's accountants' plans, namely to have Mr Walker undertake to find a third party buyer for the business as quickly as possible, and, if negotiations had come to nothing by 30th September, then the minicab business would be closed down, in accordance with a phased programme to be settled between both men. Mr Walker wrote in his own private notes at the time 'whether profit making or not!' The taxi trade was watching closely the continuing operation of the minicabs. Its trade publication 'The Steering Wheel' ran a small feature on July 7, 1962 in which an article suggested that 'we should not encourage London taxi proprietors to press for higher fares themselves at this stage' and it went on to say that 'the trade cannot be complacent, for the mini-menace is by no means resolved'.

The subject of the congested Taunton Place premises came to the fore once again, when Welbeck Motors were summoned to Marylebone Court in July for seven offences of unnecessary obstruction and one of parking in an unauthorised parking place. Pleading guilty, through their solicitor Mr N A Orman, for allowing vehicles to be parked up to five and a half hours, the company sincerely apologised for the inconvenience and said that it would endeavour to ensure that the offences took place 'a little less frequently'. It was reported that they had in fact 'got rid of a number of cars and had bought a new premises for garaging'. The company was fined £31.

Although Mr Walker had taken the necessary steps to return the company to a profitable position and he had drastically reorganised the three trading companies, (Welbeck Motors Ltd, Welbeck Motors (Car Rentals) Ltd and Welbeck Motors (Minicabs) Ltd), the Corporation was becoming anxious to extract itself from its continued involvement in Welbeck Motors as soon as possible, and, in July 1962 they told Mr Walker that it wanted to sell all of their interests, with the exception of the contract hire agreements, if he could find a suitable buyer. Mr Walker, who was considered an old and trusted friend of North Central Finance (London) Ltd, approached the Corporation with an offer to purchase the business. A surviving draft proposal, drawn up on 18 July by North Central Finance, shows an agreement to pay a purchase price of £65,000 less any contractual rights received by the Corporation, before close of business on 29 June. These rights were estimated to be in the region of £20,000 for the agreements and the finance company would loan Welbeck Motors a further £75,000 with repayments to be in equal instalments of £25,000, with the loan to be secured by a First Mortgage and General Charge on the assets of Welbeck Motors and its subsidiary companies, supported by the guarantee of the subsidiaries. It is noted within the agreement that The General Guarantee Corporation would still own the Contract Hire fleet.

Mr Walker would also be released from his undertaking to sell his shares in Welbeck Motors to the Corporation in accordance with the agreement made on 21 November 1961. In the event, correspondence suggests that North Central were agreeable to completion taking place on any working day prior to 31 August 1962. However, the Corporation, or rather Sir Isaac Wolfson, who had now been knighted, changed their position again, this time deciding to postpone the offer from North Central and instead to continue to trade Welbeck Motors and placing unspecified funds at its disposal. Mr Walker was understandably infuriated and wrote to Sir Isaac personally as Welbeck Motors was hopelessly insolvent, and it was only his own personal past association and credit standing which had persuaded North Central to make an offer, and this would have been a chance to keep the company going. Walker stated that if Wolfson was not prepared to accept North Central's offer, or to put the Welbeck Companies on the same financial footing as per their offer, it would leave him no choice but to resign. The letter was copied to Simon Coorsh, Mr Finnie, North Central Finance, the manager of Westminster Bank, Eton Branch and The Ford Motor Company.

A hand-delivered reply came though Sir Isaac's Solicitor, Paisner & Co on 27 July, in which the opening paragraphs made it clear that his letter was not a direct matter for Sir Isaac or for Great Universal Stores, but for the General Guarantee Corporation, and, even then, only in the capacity as a secured creditor. The letter finished by saying that, if Mr Walker were to resign as director, their clients had instructed them to state that they would, in all probability, appoint a receiver under the terms of their debenture.

In a board meeting of 2 August 1962, Mr Walker did present his letter of resignation, which was to take effect on the following day, along with the resignation of his son, Mr Graham Walker, to which resignations the board reluctantly agreed. This decision brought an immediate hand-delivered response from Paisner & Co and an irate phone call from Sir Isaac Wolfson himself. If Mr Walker resigned the creditors would not be paid, thus impairing Mr Walker's credit, and, of course, a receiver would be appointed immediately.

Correspondence between both parties still ignored any connection between Welbeck Motors and the Corporation. As Mr Fuller was away on holiday and as he was due to announce his intention to retire on his return, it was agreed that Mr Hodsdon should be appointed to the board as Secretary. It was further agreed that as little publicity as possible, preferably none at all, should attend

Mr Walker's resignation, as this would not be in the interests of either party.

However, Mr Walker did continue to hold his shares in the company and minutes of a meeting held in August record that, although Mr Coorsh had to record that matters had improved under Mr Walker's direction, he still painted a black picture to the remaining members of the board. In October, a Mr Sidney Goldberg was appointed manager and, in the three months under his control, the business was said to have continued to make a profit of £5,000. The car sales remained at 109 Crawford Street (tel 0561), the car hire service continued at Taunton Place, also using the same telephone number, and the minicab operation at the same address with the familiar telephone number of 4440 and of HUNter 1250. A large advertisement at the bottom of the relevant page in the telephone directory was taken out that year.

Mr Sidney Goldberg is pictured in early 1963 accepting the first of the Ford Cortina minicabs. (Mary Evans Picture Library)

21 June, 1963   CAR FLEET MANAGEMENT   *incorporated in "Motor Transport"*   Motor Transport

A CHECK-DAY FOR EACH CAR

Service face-lift for Welbeck minicabs

By JOHN R. SOUTHGATE

THREE - hundred cars, working a 24-hour day in London traffic conditions, and clocking 1,500 miles a week. How about this for a maintenance headache?

The headache belongs to Welbeck Motors (Minicabs) Ltd., the London radio - control m i n i c a b concern which hit the headlines last year through its clashes with the traditional London taxi-drivers.

Welbeck prides itself on providing a cheap cab service, but to do this it must make the maximum use of its vehicles. To the maintenance man this is a dreaded phrase, for nine times out of ten it means he comes a bad second to traffic demands, despite the ultimate importance of service.

In the past Welbeck has had its problems in this respect and its fleet of Renault Dauphines has, to put it mildly, taken a bashing. This is literally so as well, for in one year there were 75 crashes (for and against).

Now, however, Welbeck is replac-

The cost of radio-controlled minicab operation is aptly demonstrated by this pile of worn-out generators.

major clutch overhauls at 500 mile intervals.

Fifteen replacement Cortinas have so far been delivered and delivery of the remaining 285 is to be phased over the rest of the year. During a year's operation these cars are reckoned to clock 72,000 miles and at that time they will probably be disposed of, although Welbeck's policy in this respect has yet to be finalized. It seems likely that yearly replacement will be adopted, however, as experi-

are delivered they also will be allocated a check day.

The check consists of an examination of the vehicle by the garage staff (oil, tyres, batteries, etc.) and a test run of the vehicle by the workshop foreman or chargehand, who also questions the driver about performance and possible sources of trouble.

About half of the 15 cars so far concerned are checked in the morning and the rest in the afternoon, the maintenance staff leaving a list of the vehicles they want with the shift office, which administers the drivers' work.

During these checks vehicle mileage is noted by the maintenance staff so that they can gauge when they will next require the car for a full-scale service. The shift office is given three days' notice of this service to allow reallocation of drivers.

For its regular servicing Welbeck relies on the Ford service voucher scheme. This consists of a 500-mile free service (lubrication, adjustment and inspection) and is followed by more detailed 5,000-mile services.

The 5,000-mile service consists of the usual topping-up and lubrication, plus items like cleaning carburettors,

Checking and cleaning the sparking plugs are just one of many items on the list in Welbeck's 5,000-mile dock.

ignition systems and filters. Mechanical aspects of this service include blowing out and checking brake linings, adjustment of front-wheel bearings, and adjustment of valve clearances.

Service voucher booklets for each car are filed by the foreman and provide a ready servicing check. When

more Cortinas join the fleet, a progress board giving details of each car is to be compiled.

Following this system, and keeping the old Dauphines functioning, keeps seven fitters engaged full-time during the day, and three more fitters are working on a night shift.

But Welbeck feels the plan will more than pay for itself in greater on-the-road availability and reduced overhaul charges.

Pressure lubrication is carried out while the car is on the workshop lift. The lift is also used to make inspection easier.

ing its Dauphine fleet with Ford Cortinas and at the same time has taken a long, hard look at its servicing system. Both these developments stem from Welbeck's new managing director, Mr. Sydney Goldberg, who took over the minicab concern last October.

It was the maintenance angle which led primarily to the change to Cortinas in the first place. The Dauphine was not found robust enough for Welbeck's intensive minicab operation. The Cortinas, though dearer, might be more economical in the long run.

As might be expected, taxi work puts a big strain on transmissions and gear-change linkage to the rear engine, and on Dauphines gave the maintenance staff trouble. Drivers riding their clutches at traffic lights led to

Examining front-brake linings and wheel bearings is an essential point for Welbeck. Linings suffer badly in London traffic conditions.

ence with the Dauphines (average mileage now 100,000) has shown that maintenance, overhaul and repair costs are exorbitant after such a length of operation.

Even on yearly replacement, Mr. R. Packer, head of Welbeck's maintenance department, estimates the vehicles will need at least a new set of tyres (Michelin X are fitted as replacements and give up to 50,000 miles), two or three brake re-lines, and probably a new battery and generator.

With its cars doing such high mileages, Welbeck regards regular checks on them as essential, and does not rely on the drivers to report defects. With this principle in mind Welbeck has started a system of regular weekly docks with its new Cortinas.

The first batch of 15 Cortinas are scheduled to have their checks every Monday. As the rest of the Fords

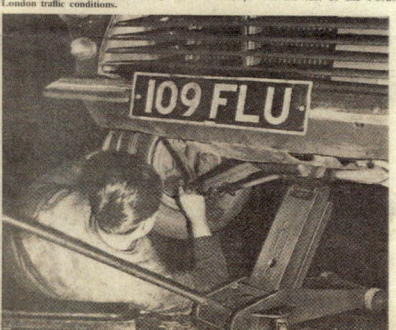

Motor Transport Magazine article, on the Ford Cortina minicab. 21 June 1963. Newspaper Advertisement. (Joel Frorath)

In 'The Times' of 29 January 1963, "Motors and Motoring" section, a story reported that Welbeck Motors' fleet of one hundred and sixty five Renault Dauphines were to be replaced with three hundred Ford Cortinas. It was recorded that the first batch of new cars, would be brought into service by the middle of the following month and, by the end of the year, the company expected to be operating Fords only. Public demand, the story continued, had continued to increase and some ten thousand orders were being handled each week.

'Motorsport' magazine in May also ran a small article about Welbeck Motors 'adoption of' the Ford Cortina for their London minicab service. It said that 'the first of these Cortinas went into service last March and within twelve months from now, Welbeck Motors expects to have three hundred of them operating a taxi service around the clock, three hundred and sixty five days a year for the people of London'. Included in the small feature, there was a photograph of Mr Goldberg receiving some of these new Fords, but, in that same month, he had become ill. Mr Walker was still a shareholder in Welbeck Motors, and on 10 June 1963, along with other shareholders, Mr E M Fuller and Mr Graham Walker, who was now representing Mill End Motors Ltd, he attended the 14th Annual General Meeting of Welbeck Motors held at the Crawford Street premises. A proposal was put forward for Mr Walker to be elected to the board and seconded by both other shareholders. Being fully aware of the General Guarantee Corporation's plans for Welbeck Motors, this was something Simon Coorsh did not want and he pointed out that Mr Walker had resigned entirely of his own volition, although he, personally, had not wished him to do so at the time. The meeting was adjourned until 1 July and Mr Hodsdon, now Director and Secretary, wrote to Mr Walker the next day in

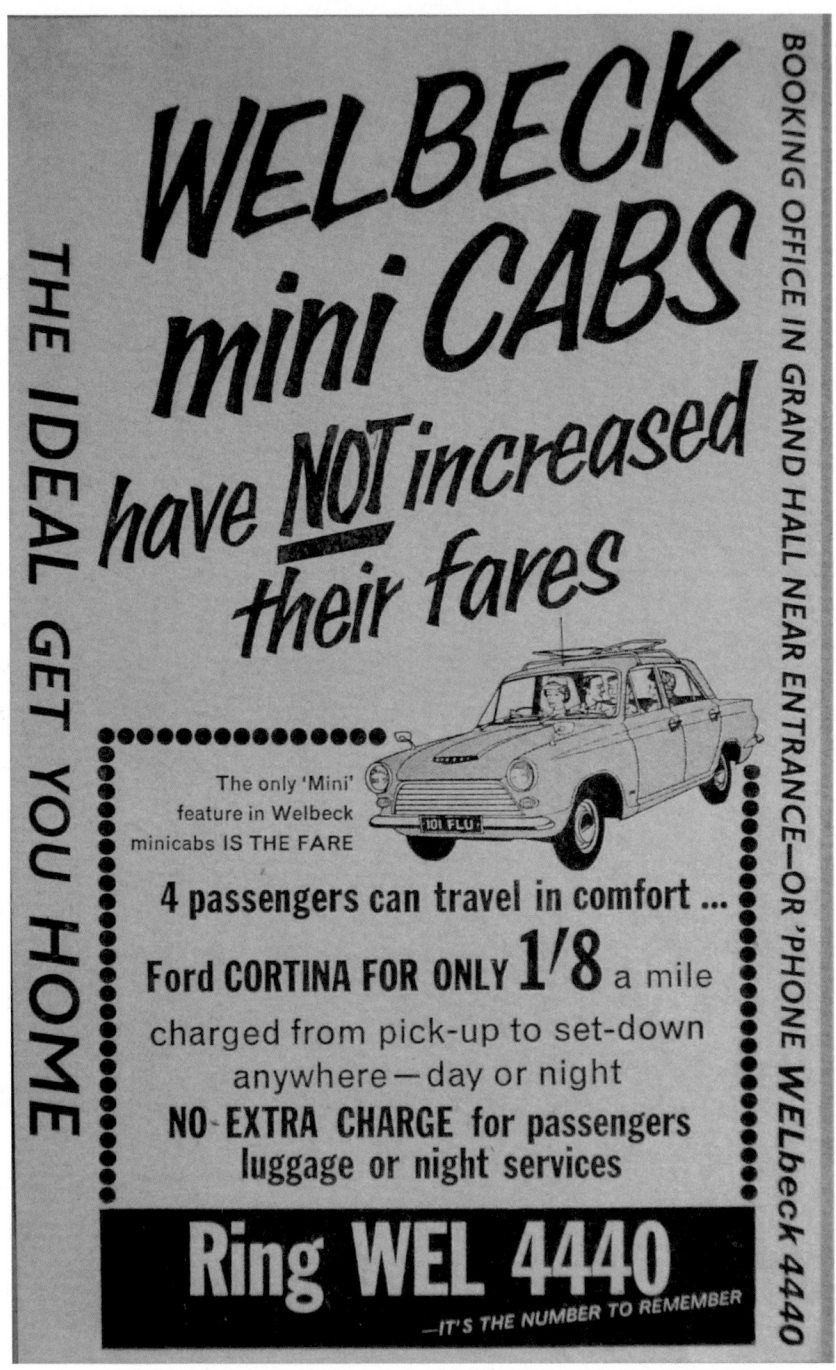

Newspaper advertisement (Joel Frorath)

order to clarify that re-election would not be in accordance with the company's articles of association. At this next meeting, the accounts for 31 December 1962 were still not present and it was decided to adjourn the meeting yet again, although, this time Mr Coorsh reported that the Corporation had decided to serve notice on the company under the terms of their debenture for a full repayment within seven days and he anticipated that a receiver and manager would be appointed shortly. Mr Walker decided that, in view of the foregoing remarks by the Chairman, he would not wish to proceed with his proposal to be re-elected.

In July, Mr Walker received a letter requesting a transfer of his shares. Hitherto, he had previously refused the request from Simon Coorsh, but he met with David Finnie in early September whereat an offer was made to pay Scot Securities (one of Mr Walker's companies), if he would transfer the shares. Otherwise, it was noted that a receiver would be appointed. Mr Walker gave the Corporation three options: it could either take a realistic price for the debenture, bid him for his shares, or they could appoint a receiver. It took until 3 January 1964 for Mr Walker to receive the payment for his shares, so that his eight years of involvement with Welbeck Motors came to a final end. Without doubt, it appears that the company's demise had come to the fore. When the company could no longer keep up the payments, the Corporation took over the company with the property and the business started to be wound down.

In the 'Evening News' and 'Star' newspapers on 27 July 1965, a small article appeared stating that a winding up order for Welbeck Motors (Minicabs) Ltd had been made in the High Court that same day. Mr Justice Pennycuick said the firm was hopelessly insolvent, with estimated liabilities of £50,000 and assets of £2,000. Petitioning creditors included the City of Westminster for £2,652 in rates arrears, supported by the Commissioners of Inland Revenue for £8,750 in tax and Welbeck Motors Ltd (the company's landlords) for £4,447. Mr Goldberg and other creditors totalling £13,295 opposed the petition. Mr Goldberg had asked for another chance to restore the firm's fortunes and he had suggested the appointment of a provisional liquidator for a trial period, with himself as a special manager. The judge dismissed this proposition as the petitioning and supporting creditors were not prepared to entertain the idea and, whilst he sympathised with Mr Goldberg, he had no alternative but to make the order. Welbeck Motors, he added, were not associated with the minicab firm. Creditors passed a resolution for the appointment of a chartered accountant, Mr T L M Macleod, as liquidator and one of the company's assets, namely the company's telephone numbers, were sold to Mr Arthur Jiggins co-director with his wife of Rushospeed Car Hire Ltd, for £700.

'The Daily Telegraph Reporter' carried an article, in which details emerged of a creditors' meeting, at which a Mr D A Williams, assistant official receiver, stated that the company had got off to a bad start when the minicab venture was launched. It had lost £100,578 in the first seven months of operating. By the end of 1961, the company was reported to have three hundred cars, but the modest hire charges had led to a demand, which could not be fulfilled, owing to driver shortages.

Difficulties with the established taxi cab drivers was also noted, together with prosecutions against the company on the grounds of operating unlicensed taxis. Welbeck Motors, it was said, were unable to find a quick way to reduce its trading losses and Mr Goldberg was then appointed Manager.

Even though the Welbeck Motors' story now finally comes to an end, the Welbeck name did carry on, due, one assumes to the purchase of the telephone numbers by Rushospeed. In 1966, the name Welbeck Minicabs was still listed prominently in the London directories, with an address at 56 Paddington Street W1 and with the familiar telephone numbers of 4440 and 1250. Further down in this publication, Welbeck Motors Ltd car sales was still shown as being at 109 Crawford Street, which was, of course, separate from the minicab venture, and there is another entry for Welbeck Motors (65) at 56 Paddington Street, with a new number of 01 – 580 2381. These trading arrangements continued into the February 1968 edition, but in July 1969 there were just three listings, two for the minicab venture now bearing the revised exchange numbers of 01-935-4440 and 01-486-1250, and one for Welbeck Motors (65) 2381, all with the address of 56 Paddington Street.

With the publication of the June 1971 edition of the telephone directory, just less than two years after the previous book, the Welbeck name had finally disappeared from the public eye.

Chapter 13

New Horizons
for Michael

After the board room differences, which had led to his resignation at Welbeck Motors and having had to walk away from the company, which he had founded, it would not be long before Michael would look to re-establish himself in the motor trade. But this time though it would be well away from the glare of the publicity media.

Contrary to the myth created by the taxi trade that Michael had more or less fled London, he had, in fact, entered into partnership as Welbeck, Knight & Long with a coach works, which was based at Dumpton Place, Chalk Farm in Gloucester Avenue NW1. Companies House information stated that it was established as Welbeck Coachworks (London) Ltd on 12 January 1962, which was less than two months after Michael's resignation from Welbeck Motors. One of the partners in this venture was Robin Frorath, who had previously worked with Michael at Welbeck Motors, having finished his National Service in 1954. Interestingly, Robin's mother had also purchased one of the new Swallow Dorettis for her son as a 21st birthday present at the time! The coach works dealt mainly in what would was considered to be 'high end' vehicles from dealerships in the Knightsbridge area, such as Rolls Royce, Bentley and Mercedes-Benz. Little is known about how long Michael remained involved with this venture, although Robin eventually bought Michael's shares, and continued successfully in business, until he sold the company in 1986 to the Lex Brooklands Group.

On 18th September 1962, Michael had registered a new company in the name of 'Michael Gotla Motors Ltd', with Yvonne Martin being the secretary of this new undertaking. After his resignation from Welbeck Motors, Michael had probably moved full-time to West Sussex as he had taken on the business of Hill Barn Garage at Sompting in West Sussex on the A27 Brighton to Portsmouth Road around this time.

This business, purchased from a Mr Albert Peffers, was ideal, as it had a detached building with ground floor showroom facilities, workshop, a flat above and forecourt parking, which faced onto a busy crossroads, this being ideal for passing trade. The BMC marque was the business mainstay at the time, with Michael said to be doing a roaring trade in selling the new Mini; and with his undertaking being dubbed 'The Mini Centre of the South' – although, sadly, many such cars came back for warranty work, due to the manufacturer's poor build quality. However, the location of the premises might not have been sheer coincidence, because doubtless, Michael would have been familiar with the area, as he had previously made regular visits there, having owned a nearby property already. On 18 December that same year, using his new headed paper, Michael wrote a general letter to his many friends and business contacts, stating that, after having resigned from Welbeck Motors due to a disagreement with Wolfson on the future of the minicabs, together with Mr Walker and most of the old Welbeck staff, he had bought Hill Barn three months previously and he invited the recipients of the letter to call in, if ever they were in the area. At the bottom of the letter, the Secretary of Hill Barn was now noted as being Stanley Ardley.

Although Michael was now no longer a subject of particular interest to the national newspapers, a small piece did appear in one such paper, which was headed 'Minicab Fleet Takeover' and in which it was stated that Welbeck Motors was being taken over by a company by the name of Red & White Cabbies of North London, and that it would manage the operation for a fee. Whether this was actually the case or whether it was just creative reporting shall be left for the reader of this book to decide, but the article did state that Michael had announced the building up of a new business on the South Coast. The article went on to indicate that Michael had bought Hill Barn Garage for around £25,000 (again it is likely this figure may not have been correct) and that he hoped to buy several more businesses in the area. It stated further that, this time, Mr Gotla was working alone in the new venture and that he had no plans to go into the minicab business again.

Where the business was located was said to have been the site of an even older, pre-war garage premises, although the Hill Barn trading name does not appear in the local telephone directory until the May 1959, when it is first listed as Hill Barn Motor Co, 75 Upper Brighton Road. (Telephone 3085). By May 1963, an additional line (4618) had been installed. The detached premises featured a showroom, which could house between three and four cars at ground floor level, with Michael's office at the rear of the open space, looking out onto the forecourt, with the sales manager's office next door, which had access to the workshop at the rear. There were operational petrol pumps on the large forecourt area with attendant service and Michael, himself, had occupation of the flat above, choosing to live on site. He decided to have a large canopy fitted round the front of the building to give it the appearance that the property was much larger than it actually was, he had ramps installed for ease of maintenance and he updated it generally. On occasions, prizes were awarded to the garage by the fuel suppliers for the best forecourt and garden, such was Michael's attention to detail on the presentation of the premises, with pictures being published in the local newspapers.

Thompsons Garage at Sompting is seen before its change of name in the 1950's to Hill Barn Garage – (Lancing & Sompting Pastfinders)

The business itself was much more manageable than the previous Welbeck operation, with a total staff of a workshop manager, a salesman, a chief mechanic and his assistant mechanic, a trainee mechanic and a vehicle preparation person. There were support staff for the administration and the business thrived on the sales of new BMC products including Austin, Morris, Riley and Wolseley, as well as all the popular versions of the Mini, including the Countryman and the Cooper versions.

Hill Barn employee Roz tending the forecourt garden at Hill Barn Garage – Michael Gotla Motors. (Roz Horner)

Vehicles were collected from the local distributors, Caffyns in Brighton for the Morris & MG marques, and from Wadham Stringer for the Austin brand. These vehicles were brought back for de-waxing and for preparation before sale. Occasionally, late models of second hand vehicles were sold alongside new products, but trade-ins were normally put out again through the trade.

Other premises followed. The Budget Shop was established in nearby Lancing for second-hand car and petrol sales and it was run by Albert Peffers, from whom Michael had bought Hill Barn. There was also Brighton Road Garage on the Lower Brighton Road; and Newhaven Garage in the High Street just over the county border in the East Sussex port town of Newhaven, which was run by a Mr Brian Heath. These latter premises suffered a major fire in 1976, when trading as Arrow Tyres, long after Michael's ownership.

Michael would have had the use of new cars via the dealership and one such sports car, an MGC was driven by him to the island of Gozo to visit friends for two weeks – quite a journey! It was not unknown for Michael to take a trip over to Europe and he would come back with a car. One such vehicle, as recalled by a member of his staff, was a lovely white Alfa Romeo Guilietta, with a five- speed gearbox.

It was only in the May 1965 edition of the telephone book that Michael Gotla Motors Ltd is finally in print in the directory. An Austin A40 service van (and latterly a new grey Mini Van) would carry his name on their sides. Like many relatively small garages, a company Land Rover was always available for breakdowns and towing duties when needed. Michael is remembered by surviving staff as taking a genuine interest in them, generally talking to them in a relaxed manner and looking after them. He would host parties in his flat above the business, with customers invited, too! The local

Michael on Gozo c1969 on holiday touring in an MG. (Roz Horner)

pub at this location was set back from the main junction across the road and the 'Joyful Whippet' was certainly not unknown to Michael, who would pop across regularly for refreshment at his local.

Michael's entrepreneurial eye could not resist the temptation of buying a property for renovation around June 1968. It was named 'The Hermitage' just down the road from Hill Barn in Church Lane, Sompting. Michael still used to travel up to London, where

HILL BARN GARAGE
MICHAEL GOTLA MOTORS LTD

Morris Sales and Service

The Mini Centre of the South Coast

There is still time to have a new Mini on the road for Easter

Lowest Interest Hire Purchase   Best Possible Exchange Prices

USED B.M.C. CARS

1963 (January) MORRIS MINOR 4-door De Luxe Saloon. Mileage only 700 (quoted last week at a different figure by a mis-print), originally supplied by us and there is a very genuine and sad reason for sale. Quite as new, rose taupe, taxed. Would cost new £580 .... £520

1962 AUSTIN MINI De Luxe Saloon. Yellow, mileage 12,000, as new, owned by local school-teacher £425

1955 (Registered November 1954) MORRIS MINOR Convertible. This car is in quite superb condition and is well worth the high price we are asking £215

1955 WOLSELEY 4/44. Dark green, mileage 47,000, possibly the best in existence £275

These non-B.M.C. cars recently taken in part exchange are offered at exceptionally low prices

1961 (Registered November 1960) FORD POPULAR De Luxe. White, mileage 7,000, as new £340

1958 FORD PREFECT De Luxe. Beige, a specimen car £275

1960 BEDFORD UTILABRAKE. The de luxe two-tone model with all extras, 12-seater, reconditioned engine recently fitted (registered December 1959) £335

1956 BEDFORD 10/12-cwt .VAN. Identical to much later model, fitted with windows and seats. M.O.T. tested £100

1953 FORD 8 h.p. UTILICON ESTATE CAR. In mint condition, owned by County Council .. £85

1962 TRIUMPH HERALD 1200. Mileage 19,000, blue, disc brakes, very slight damage at rear. For quick sale £450

HILL BARN GARAGE
MICHAEL GOTLA MOTORS LTD

75 UPPER BRIGHTON ROAD, NORTH LANCING
LANCING 3085 and 4618

On the main A27 Upper Brighton to Worthing Road, about eight miles from Brighton, two miles from Worthing

Telephone us and we will gladly bring any car to your home for inspection and trial without obligation

Newspaper advertisement for Hill Barn Garage sales. (Authors collection)

his friends were, and he had a business involvement with Mr Jimmy Gregory of Kingston Hill Motors (famous for being the owner of Queens Park Rangers Football Club), whilst he built up a good motor dealership in Sussex.

Michael sold HBG around 1967-8 to National Benzole and leased it back, but in 1970 he re-assigned the Hill Barn business lease to Mr Andy Elwin, a former racing driver (and winner of the Television Trophy) and, around this time, he had taken on a further move by purchasing a large property at 30 Brunswick Terrace in nearby Hove for around £17,000. Michael's intention was to convert the property into flats, but this turned into a costly project, due to the nature of the work needed.

However, within two years, Michael had decided to move to Ireland and, in a letter to a friend, he commented on how he was living very quietly at the time. He did still return to England on occasions, notably for the twenty-first Birthday of his brother's youngest son. He was living at a property called 'Broc House' in Dublin, but the lure of business could not be resisted and, on 9 August 1976, Michael took over a very old established Day Cleaning, Launderette and Laundry business in Dun Laoghaire, previously Kingstown, a suburb of Dublin. Michael felt that the business was well positioned in the main street and it was only a couple of miles' walk from the sea. Like Hill Barn before, it came with a large flat, which he felt might be gutted and made into a good home for him. He commented that 'the old team are all back in action' as Lombard and Ulster Finance, (the Irish Republic's version of North Central Finance) were putting up a lot of money for the business.

At the time, Michael felt that he might stay over in Ireland for around five years or so, and, at times, he felt a little homesick, although he commented on the fact that, strangely enough, when he came over to England, he was glad to get back to Ireland. He said that it was all so easy-going and very different in Ireland from the regimented English attitude. It amused Michael when he read in the press that the Minister of Transport had been asked a question in the Dail (The Irish Parliament) as to whether he was going to make seat belts compulsory in cars, as in all other Common Market countries? The answer was 'Certainly not, what would be the good? The Irish motorist wouldn't take the slightest notice'. Michael also noted that neither MOT testing nor the wearing of crash helmets existed in Ireland. But, above all, everyone was so friendly and strangers talked to each other, whilst shop assistants would go to immense trouble for a customer, yet nobody appeared to work very hard!

Michael in a relaxed pose with nephew Peter Gotla c 1959. (Peter Gotla)

Eventually Michael sold the Dry Cleaning business and, ever the entrepreneur, he decided to buy two racehorses in foal, but, tragically, both died thus losing Michael most of his capital in the process. By April 1988 Michael had arrived back in the UK and had settled in a flat in London Road, Leicester in the town where his brother, Dudley lived. He was now described as a retired company director. By July 1994, Michael had moved again, to a flat at Riverside Court in Market Harborough before one final move to a flat at 'The Old Vicarage' in St James Road, Leicester. Eventually he passed away there on 29 August 2002 at the age of eighty three.

When Michael's flat was cleared of his possessions by his relatives following his death, a small mint condition Dinky Toys' model Renault Dauphine minicab was discovered, amongst other motoring memorabilia, perhaps being one last poignant reminder of a lost time in a life, which had now come to an end with the death of Michael Gotla.

Chapter 14

Looking Back

When one has the benefit of hindsight to review a particular period in time, it is all too easy to see the struggles, which took place. These struggles showed clearly that the Welbeck Minicab venture could not have sustained itself in view of the outside factors, which were beyond the company's control.

Undoubtedly, there were a number of factors, which prevented Welbeck Motors' new minicab venture from succeeding. Opposition from the taxi trade was a constant physical problem, which caused much disruption and which did not assist the operation to run smoothly. One has to sympathise with the fact that a traditional 'cabbie' was obliged to pass stringent tests and to have learnt a vast amount of intimate knowledge about the streets of the capital, before even being able to become fully licensed to drive a taxi. This was (and still is) a real skill which requires a person with a certain level of professionalism to be able to undertake. On the other hand, a minicab driver, due to the way in which the law was framed, could undertake his role and perform the same basic task, albeit without the same intimate knowledge of the streets of the capital. Indeed, his present day counterpart has the additional aid of an extremely accurate and simple Global Positioning System, such as Tom Tom, often built into his vehicle!

Minicab firms had difficulty in finding enough drivers, let alone ones with enough basic knowledge of the capital to be suitable for the position, and this was a huge stumbling block to the new operations. Being able to expand fast enough to meet the unexpectedly high demand and to bring in sufficient income to generate the necessary profit was equally hard. Carline had carried on a brisk trade with their uniformed drivers from the outset of operation, but, early on, of the one hundred and eighty-one potential drivers interviewed in one week alone, only three candidates were suitable to be taken on. Any business, which attempts to provide a service at a much lower price than the current market level, can only achieve success by operating on a large volume.

In Welbeck Motors' case, many a taxi owner of long experience was convinced that, with the low tariff of one shilling a mile, the new experiment would fail and, ultimately, they were proved to be right. The eventual loss of revenue from advertisements adorning the Welbeck Minicabs can be traced to the very public opposition to this innovation by the taxi trade and through the national papers and media. Such commentary created a very negative effect with the companies, which had taken various advert sections on the vehicles, thus forcing some to pull out. All this, combined with the final court ruling that even the remaining wording of Welbeck's own name and telephone number had to be removed from the minicabs, ensured that the income from this novel idea, designed to offset much of the running costs and to subsidise the fare structure, meant that the real potential of the idea of 'project minicab' could never be realised.

The General Guarantee Corporation's eventual decision to withdraw their financial backing of this venture, into which they had poured so much money, was, to say the least, regrettable, although understandable. Whilst unforeseen and uncontrollable changes in the fortunes of the national economy squeezed all the Welbeck companies at the time, there was actually much further potential growth and profitability for the business both in car rentals and sales as well as in the minicab venture. Welbeck Motors was by no means the first company to venture into the 'new world' of minicabs, and, like many of the other firms which had set up and which have now become part of history, that point is generally overlooked. But the Welbeck name still conjures up particular images for those of a certain age. Taxi men may still remember it as a threat to their livelihoods, which demanded the drastic action which the trade took to ensure that this new and unregulated type of operation would not succeed. For the public however, the image was of a less expensive form of transport which offered a 'go anywhere, go any time' service. Minicabs were much needed to fill something, which one would now call 'a gap in the market', and which was not being met adequately by the established taxi trade of the time. Indeed, the fact that the minicabs were clearly in such high demand suggests that this need for minicabs was more than slightly true.

As the public face of Welbeck Minicabs, Michael had come to know many men of the press in his time and, judging by the long list of names, which he kept in his diary, there would generally have been a mutual bond of trust between them. With his background flair for advertising, Michael knew that communication was key in any industry dealing with the public and there was scarcely an article amongst the hundreds printed, which portrayed the venture in a bad light. However, what was almost incredible was the publicity behind Welbeck Motors' efforts to break into this new market. Its own impressive advertising campaign was totally eclipsed by the huge, and not always intentional, amount of newspaper editorial. Michael and his management team were interviewed on radio and on television, and, together with questions in Parliament, it was perhaps, as one newspaper suggested, *the* story of 1961.

In a letter to 'The Observer' newspaper on 21 September 1961, Michael wrote the following, which seem appropriate from a forward thinking man, who, as the reader has hopefully deduced, was ahead of his time: 'There is nothing wrong with the London taxi driver and nothing much wrong with the regulations governing him. But the specially designed London taxi is a complete anachronism. The demand for this vehicle is so small that soon no manufacturer will want to make it. Minicabs might not sweep across London, but the production car is bound to be the London taxi of the future. In their desire to retain this funny old vehicle, the London taxi leaders are out of step with the whole world'. Michael was characterised by his nephew, Andrew as a real showman, always having a selection of cars available to drive, usually including a 'flashy convertible', and above all he was 'a real man of the motor trade'.

In my research a number of minicab drivers names have been recorded through various press cuttings and documents which I felt needed paying mention within these pages, if you, the reader, should recognise any of them as family members or friends I would very much like to hear from you, or indeed if you know of anyone with a Welbeck Motors connection.

Minicab drivers are filmed walking towards the news reel camera by
British Pathé on the launch day event on 19th June 1961. (British Pathé)

Of the fifteen ladies initially said to have started when operations
began, the few names recorded were; Jane Simpson, Vivien Batchelor,
Kathleen Edwards, Betty Puri and Avril Strong.

Of the many male drivers employed, the names recorded were;
Oliver Davis, Arthur Pemberton, Phillip Frank, Robert Wynn,
John Hopkins, James Vincent, Eugene Cahill, James Yates, Aldon
Brysen-Harvey, Frederick Jones, Stephen Wojcik, Dennis Goodwell,
Arthur Clements, Mr Turner, Derek Vincent, Eugene Howard,
Edmund Courtenay, Alan Barraud, Mr T. Jirapure, James Firth,
John White, Arthur Morecroft, Robert Johnson, Thomas Rivers,
Albert Beeland, Ted Cantwell and John Stanton.

Appendix

Traces of Welbeck Motors and Michael Gotla's legacy.

After detailed research through the Driver and Vehicle Licensing Agency (DVLA) by the author, none of the well-known fleet of Renault Dauphine minicabs appear, unsurprisingly, to have survived, with one possible vague exception.

One sole Renault did actually meet the final 30th November 1983 computerisation deadline of the then Driver and Vehicle Licensing Centre (DVLC) for older vehicles in the UK requiring their older style logbooks to be upgraded to the then newer V5 document. For this one survivor it was issued on 26th October 1983 when the car would have been some twenty-two years of age at the time. One will leave the reader to decide on its possible later fate.

Searches of the known number plate allocations for the later Ford Cortina's have also not shown any survivors of this model either. By the nature of the vehicles commercial use when new as minicabs it would have certainly given these cars a much lesser chance of survival in the longer term.

The former Welbeck Motors buildings at both 95 and 109-111 Crawford Street still exist many decades after the company's demise, albeit now in a different commercial use. But, if one stands in Crawford Street and surveys the local scene, it is not hard to imagine the activity, where one of the big players in the London automotive scene once held court. At one time the building at 109-111 was occupied by a design company, which was responsible for the Anglo-French train, 'Eurostar'.

However, the property at Taunton Place has been demolished long since, in order to make way for modern flats, thus taking with it the legacy of a brief period in the early 1960s, when it was the hub of the country's then largest minicab venture, which was of interest to the public and to the press alike.

The former showrooms in Crawford Street, London 2019. (Stephen Dine)

The Boston Arms Public House and adjoining former Welbeck Motors Garage are no longer part of the Taunton Place scene in London, replaced by a much larger building. (Stephen Dine)

Altan Motors, formerly Hill Barn Garage, in 2016. (Stephen Dine)

In December 2018 Joel Frorath (pictured) meets the
author and presents him with a Welbeck drivers cap kept
by his late father Robin Frorath. (Stephen Dine)

Close up images of the surviving Welbeck Motors
Minicab drivers cap (Stephen Dine)

The premises of the former Hill Barn Garage, home to Michael Gotla Motors Ltd, did still exist into the 21st Century looking out across the now busy A27 dual carriageway between Brighton and Worthing. It was still in use by the motor trade up until 2017, latterly as the premises of Atlan Motors, and was in a very little altered state from Michael's time there in the 1960s. In late 2019 the property now stood empty with an uncertain future and sadly in January 2020 it was finally demolished to make way for a new development.

Across the road, Michaels' former local haunt, 'The Joyful Whippet' public house was still in business in early 2020, albeit having had an application for demolition and development put forward, although in the March, due to the Global Pandemic, it would become permanently closed.

Comparisons of the original and reproduction minicab models together (Stephen Dine)

The 2018 De Agostini brochure front cover illustrated with historical facts about the Renault Dauphine minicab that accompanied the purchase of a minicab model as part of the Dinky Toys classic collection (De Agostini)

In 2018 the original Dinky Toys Renault Dauphine minicab model was revived by manufacturer De Agostini, a parent company of Atlas Editions for release as part of their new Classic Dinky Toys collection, this model being their sixteenth release in the series, with the only slight difference on the 21st Century minicab model now having adverts on both sides for Kenwood, as it was not possible to use under license the original Meccano name for this very convincing reproduction model.

Memories from Hugh Ruding Bryan – Salesman at Welbeck Motors 1950s–1962

'I recall in the early days, when running Welbeck Motors, Michael had a reputation as a 'terrible gambler' and one night, in a particular card session, he had gambled (and lost) a brand new Austin A40 playing poker! The following day the company's change in circumstances and the loss of the car became painfully apparent, when the successful gentleman card player came to collect the vehicle and Michael was in a foul mood for weeks. Tactfully, I enquired of Michael, 'Mr Gotla, but one of our cars has gone' to which Michael replied 'Don't worry, don't worry. I'll tell the insurance company it's been written off!'

At the time, it was not unusual for me to be sent all over the country, after Michael had spotted a classified advertisement, in order to purchase stock for sale, such as Armstrong Siddeleys and Jowett Javelins or Jupiters, and, later on, Volkswagen Beetles. It was my responsibility to check over the car being purchased and to ensure that it was in good overall shape. Early on, I didn't know much about the motor trade – but I soon learnt! The trouble I would get into if hidden bodges had been made to a motor vehicle, meaning that they had to be rectified, soon speeded up this education.

Armstrong-Siddeleys were always a popular make in car sales and, when I was twenty one years of age, I would collect new examples from The Lancaster Group of Wimpole Street in London. These cars were stored until collection in a basement at their large premises. The large cars, with most having pre-select gearboxes, were certainly a quality product at the time and on one occasion, unfortunately, I managed to scrape all down one side of a new car upon leaving the basement of the premises. On arrival back at Welbeck's showroom, Michael saw it and said, in a few choice words, for me to go home and not to come back! I didn't go back for three days. Eventually Michael said, 'OK, you can come back now, but you have to pay for the damage to the car'. With Michael being of mixed Indian and Irish descent, he could fly into a temper, which was more often than not!

The sum of £2 per week was deducted from my pay for many months after afterwards as the damaged car needed its wing and bumper to be straightened, as well as the substantial paintwork damage to be repaired.

As well as the new cars being taken into stock, good quality second-hand examples would be sourced for preparation and for re-sale. A particular vehicle of note, which came in was a car thought to be known as one of the 'Docker Daimlers'. I had been sent off by Michael to collect it with the instruction, 'Don't come back unless you get it for a price I have told you' It was successfully purchased from its first owner, Lady Docker, and I recall that it was a one-off type, especially made for her by Daimler. A green convertible two seat with dickey seat, it didn't take long to sell. There was a large profit in it, which 'at that time' was unheard within the company. Bearing in mind that I was earning something in the region of £10 per week, the sale ensured for me a promotion within Welbeck Motors, too.[*]

Michael was an ideas man. As well as writing all his own copy for the many varied and regular advertising pieces carried in magazine publications, his guaranteed repurchase scheme could be said to be the forerunner of the hire purchase system now. Michael's saying was 'Everything has a price, never forget it. I don't care what anybody says, everything is worth something'. For example, a member of the public might come into the showroom and enquire how much their car was worth for the company to buy it. A reply would come back of say £300, with the customer saying 'Don't be ridiculous, it's worth £500.' 'Yes' I would say, 'but you're asking too much, which is why you can't sell it and you have come in to us. It's actually worth £300 not £500!' Michael's saying about everything having a price was so right, and Welbeck Motors, like many motor traders, would buy most vehicles, which came through the door, if the price was right. It was not uncommon in the motor trade that, when a good clean example of a second-hand purchase or exchange came in for re-sale, the notion of reversing the speedometer on occasions could

---

[*] Authors note: although unable to verify, the car described by Hugh may well have been the Daimler show car on display at the 1948 Earls Court exhibition, the first London Motor Show since 1938. Its 'Special Sports' drophead coupe coachwork was bodied by Barker and noted at the time as a 'an exceptionally attractive design with a modern appearance but with beautifully balanced proportions and graceful curves'

Features included a speedometer trip that could be switched to work in miles or kilometers, triple windscreen wipers and a powered hood. Priced at £7,001 it was the most expensive car at the show, gaining the nickname 'Green Goddess' due to its greeny-turquoise shade of colour.

be considered, in order to enhance a car's appeal to a prospective purchaser. One of Welbeck Motors' car valets and general hands, Burt Carter, had been instructed to take a speedo out of a nice condition car, which showed 40,000 miles with the proviso that a reverse to 20,000 would look genuine, due to its condition. Such a task was all too easy on vehicles of the time. Fits of laughter issued from under the dashboard, which collected an audience, as Burt produced a hand-written note, which had dropped out from behind the car's speedo and which read 'Oh no – not again!'

When the new showroom was eventually built at 107 Crawford Street, becoming in the process 109–111 (sometimes referred to in later correspondence as 107–109) with the adjoining vacant site next door, Michael had a flat above this property. In this, he had installed the loudest hi-fi equipment, which you had ever heard! Apparently, the whole of the street would shake when Michael had it turned on in the mornings. The services of a Mrs Parker were employed and she would come in and prepare Michael's breakfast, as well as do general housekeeping duties. Michael would normally work on his advertising paperwork upstairs before he got up, but he would sometimes forget something and he would come down into the showroom itself at mid-morning, still in dressing gown and slippers, hair all untidy and with a cigarette in mouth shouting 'Parker – where's my breakfast?' The startled look of customers just contemplating buying a car in the showroom and thinking 'Who is this man?' must have been a sight to behold. He was just a character like that, that was Michael Gotla. Michael had what was described as the loudest laugh you ever heard, and in any restaurant in London he was known for holding an audience, just when eating. Food all over the place, drinking wine, getting drunk! He was just one of those very loud people. It would turn heads around him. He was later dubbed by the press as having 'restless energy and a lively imagination'.

A regular customer at Welbeck Motors was Doctor Rossdale, a female doctor, who lived just around the corner from the showroom in Grosvenor Square. She had been supplied with the first of the new Ford Prefects and had run them for years. All of them had three-speed gearboxes with overhead valve engines, and they were nice little runabouts. When the new Ford Anglia was launched, it was considered to be much more advanced than the earlier cars. Doctor Rossdale was delighted with her new example, when she made her next purchase, but, within weeks she came back to the showroom with a query. She said that she was delighted with the car, but that it really was very, very bad on petrol. I asked what was wrong with it and a consumption figure of twenty miles per gallon was quoted. Nothing obviously wrong was found with the car, so I suggested that we should go for a drive 'Just to check out her driving'. It soon became apparent that the doctor only got up into third gear. The new Ford Anglia had a four-speed gearbox, but, after a succession of three-speed Prefects, she didn't know the new car had another gear!

Memories from Tony Brooks – Apprentice mechanic at Hill Barn Garage 1963–1966.

I first started out at Hill Barn in 1963, after seeing an advertisement for an apprentice mechanic in the local 'Argus' newspaper.

I recall that my boss, Michael Gotla was always immaculately turned out, in shining shoes and a smart suit, he could have passed for a film star. He could be volatile on occasions; he would suddenly blow up over something, but then he would calm down just as quickly. Interestingly, I never heard him swear, and I never had any problems with that side of him.

I recall that Michael was able to obtain one of the first new Mini Mokes in the area for sale. Upon the vehicle's arrival, he said to me, fill it up with petrol, go and get your friends and take it out into Brighton and the local area. He gave me a stack of brochures for the new Moke, so that when people looked at it when we parked up and were on the move, Michael said give them a brochure and tell them to come and see us about buying one!

We also had one of the first new Austin 1800 'Landcrabs' in our area, and, after preparation, it was taken into the showroom to be put on display for sale. The following morning we came in to find it had leaked oil all over the floor! In the end the problem, first thought to be an oil seal, in fact turned out to be a porous cast engine block.

Memories of Roz Ford (née Horner) – Secretary at Hill Barn Garage 1966–1986

I first saw the job advertisement for a position at Hill Barn in the local paper and was drawn by the fact that it read 'Wanted – unusual woman'. Out of the many applicants who came forward, Michael chose me, even though he could have interviewed many more.

One of Michael's sayings was 'Any publicity is good publicity' and he would always remember people, as he had a very retentive memory.

Michael had bought Hill Barn from a lovely fellow called Albert Peffers. We had four garages: Albert ran 'The Budget Shop' for Michael in South Street, Lancing, which offered petrol and car sales. There was also a lovely showroom on the Lower Brighton

Road (A259) Lancing, with a flat over the top, which was left empty, until we relinquished the lease. At the same time, we had the Newhaven Garage and of course Hill Barn, all of which were leasehold except the latter.

The day after I joined HBG in 1966, John Hamblett (who was Michael's co-director) was killed in a car accident on the A27 by Shoreham Airport. This was a terrible blow to Michael, as John was his protégée and left a widow and two children. At HBG, we also had the MG franchise, which, when the Mini craze cooled, off kept us going. In about 1967 he made me company secretary, which was an honour for someone aged only twenty-two.

Whilst owning Hill Barn and the associated garages, Michael would still go back up to London and work with Mr Jimmy Gregory of Kingston Hill Motors, keeping his hand in up there in the motor trade, often coming back down at the weekend with high-end cars, such as an Aston Martin or a Rolls Royce, or sometimes trading cars from Hill Barn up there. Other trade friends included Bob Borrodale, MD of Endeavour Motor Co Ltd, Tommy Sopwith Jnr, from Brighton Ford Main dealers, and Guy Salmon of Guy Salmon Motors of Thames Ditton.

I was told that when he started the minicabs, he had a Porsche with the number plate TAX 1, and this is when he got cornered by London cabbies and attacked, as was reported in the national press.

He still read a magazine called 'St Pauline', which I later learnt was associated with his early schooling, and he always paid into the Benevolent Fund for Motor Traders, although in the later years of his life he never asked for any help from them.

Michael sold HBG around 1967-8 to National Benzole and leased it back, but in 1970 he re-assigned the lease to Andy Elwin, who had previously worked in Forest Green, racing and preparing Minis. Andy had previously won the Television Trophy for saloon cars, presented by Murray Walker, which event was televised.

Michael and I kept in touch many years after Hill Barn and he would often write, which he liked doing.

Memories of Mervyn J Thomas – Resident of Crawford Street, London from 1920 until 1949

I was born in Crawford Street, on the corner of Upper Montagu Street in 1920, above my father's pharmacy 'Meacher Higgins & Thomas' (established in 1814). Soon after World War Two, we noticed an empty shop further down the road, which was being decorated and later there was a sign saying 'Welbeck Motors'. Soon there were several large pre-war cars parked around there. It was a chauffeur-driven hire company and it seemed to be very busy. The proprietor was a tall, handsome man, who often came into the chemist and was very friendly. He was usually accompanied by a big dog, I think it was a Great Dane.

Letters began to appear in the local newspaper complaining about the many cars parked in the street causing traffic problems, (no yellow lines or meter maids in those days).

I was married in 1949 and moved down to Worthing in Sussex soon after to work at Shoreham Airport for R G Miles. Amongst the many projects we had, there was a contract from the Road Research Laboratory to make a vehicle to test the various road surfaces using a Jaguar car. One day, we needed a wheel and tyre. I drove to the nearest BMC garage and, at Hill Barn, I was surprised to be greeted by 'Hello Mervyn'. I looked round and there was Michael Gotla! I got the impression that things had become too complicated for him in London. Small world eh?

Memories of Graham Hawkins – Hill Barn Garage

I first started out in my career by working on TVs, but I didn't fancy the job, so I started with Wall Bros, who were specialists in Rolls-Royce & Bentley cars. We used to have Norman Wisdom's car in (and I was told to stay out of the way!). I remember that he had blocks on his foot pedals. When Walls sold up, I didn't like the new owners so I went up to Hill Barn, as I knew the owner, Albert Peffers, and, of course, later, Michael Gotla took over. As a young seventeen to eighteen year old, I was not interested in politics or the news, so it didn't interest me (Michael's minicabs, before Hill Barn). You were just in a job, he was the guv'nor and I didn't take any notice of that.

Mike used to do the advert for Lotus cars; he was a very clever and intelligent man. He made Hill Barn the Mini centre of the South. We had a special trailer built, which would carry two Minis in tandem, we used to run about in a little Minivan, which would pull them; the trailer was twice the length of the van! We would go all over the place to collect Minis for selling on, and, although Mike was not a main agent, he would sell a lot of Minis, as well as Austin 1100s and MGBs. At one time, we had a yellow MGB parked on our forecourt on the grassed area with hundreds of daffodils around it. It had come with chrome wire wheels and it was sold to a jockey. (Both Fontwell and Goodwood racecourses in West Sussex were not far away).

Mike would say 'I want you to come up to London tonight.' and I would reply, 'Where to?' He would just say 'Don't worry about that, you just follow me!' In the City, once the firm's Minivan trailer was loaded with cars, Mike would say 'Off you go then' (for the return journey) and he would stay over in London and come back the next day.

Graham Hawkins behind the wheel of the Hill Barn Garage Land Rover,
when taking part in a local carnival in the 1960's. (Graham Hawkins)

Mike didn't have personal transport, he just took any car, which came in and was taxed. He did buy a Lotus Cortina. What a cracker of a car! With the 1500cc engine in it, he would go round a corner on three wheels because the suspension on it was so stiff! White in colour with green stripes, it was his personal car for a few days. Mike didn't really entertain specifically in his flat above, and, as far as one can remember, a lot of the time we were all up there having a party anyway! When I first got married, Mike bought me a fridge. He was very generous. I felt that Michael was the sort of person, who would only stick at something for so long then got bored with it. Then he would look at things and say what can I do now.

Research

Interviews with Graham Walker and Hugh Ruding Bryan, formerly of Welbeck Motors.

Interviews with Roz Ford, Roger Ashburner, Graham Hawkins & Tony Brooks, formerly of Hill Barn Garage.

Interviews with Joel Frorath, son of Robin Frorath.

Telephone Interview with Andrew Gotla.

Internal documents and papers from Welbeck Motors Ltd files 1956 – 1963

The Steering Wheel 'Journal of the British Taxi Industry' Volume LXX 18 March 1961 to Volume LXXII 20th January 1962.

A Century of London Taxis – Bill Munro

Crawfords Public Relations Division (1960)

Durrant's Press Cuttings (1960 – 1962)

Which included the following newspapers:

NATIONALS & OTHER LONDON PAPERS

Balham News, Brentford & Chiswick Times, City Press, Clapham Observer, Courier & Advertiser, Daily Express, Daily Herald, The Daily Mail, Daily Mirror, Daily Record & Mail, Daily Sketch,

The Daily Telegraph, Daily Worker, Eastern Daily Press, The Economist, Evening Standard,

Evening News, Evening News – Night Special, The Evening News and Star, The Financial Times,

The Guardian, Guardian – Journal, Hackney Gazette, Hampstead News,

Hampstead & Highgate Express, Hornsey Journal, The Independent, Illustrated London News,

Islington Gazette, London, Kentish Independent, Woolwich, Kentish Mercury, Greenwich,

Kilburn Times, Marylebone Chronicle, The Morning Advertiser, National Newsagent,

The New Daily, New Statesman (London Diary), News of the World, North London Press,

Norwood News, The Observer, The People, Punch, Reynolds News, South London Observer,

South London Press, South Western Star, Clapham, Marylebone Chronicle, Marylebone Mercury, St Marylebone Record, The Morning Advertiser, North London Press, St Pancras Chronicle,

Scottish Daily Express, Fleet St, London & Glasgow, Sphere, The Spectator, Statist,

Streatham News, The Sunday Times, Sunday Telegraph, The Tablet, Tatler, The Tatler & Bystander, The Times, Topic, Tooting Gazette, Tottenham & Edmonton Herald, The Tribune, Vogue,

Walthamstow Guardian, Walthamstow Post, West London Observer,

West London & Fulham Gazette, West London Press, Chelsea, Westminster & Pimlico News, What's On, Willesden Chronicle, Wimbledon Borough News.

PROVINCIALS & OTHER PUBLICATIONS

Advertisers Weekly, Bath Weekly Chronicle & Herald, Bath & Wilts Evening Chronicle,

Bexley Heath Observer, Birmingham Evening Dispatch, Birmingham Mail, Birmingham Post,

Bolton Evening News, Bournemouth Evening Echo, Bournemouth Times & Directory,

Brentford & Chiswick Times, Brighton Evening Argus, Sussex, Bristol Evening Post,

Bristol Evening World, Caernarvon & Denbigh Herald, Cambridge Daily News, Catholic Times

CCF News, Chatham News, Kent, Chatham Observer, Kent, Chatham Standard, Kent

Christian Herald, Chronicle and Echo, Northampton, Church of England Newspaper

Courier and Advertiser, Dundee, Coventry Evening Telegraph, Crompton & Royton Chronicle,

Croydon Advertiser, Croydon Times, Daily Record & Mail, Glasgow, Dartford Chronicle, Kent

The Diplomatist, Dumfries Standard, Dumfries, East Anglian Daily Times,

Eastern Daily Press, Norfolk, Edinburgh Evening News, Eltham and Kentish Times, Sidcup Hill

Erith Observer, Epsom & Ewell Herald, Evening Chronicle, Manchester Evening Chronicle,

Newcastle-on-Tyne, Evening Citizen, Glasgow, Evening Dispatch, Edinburgh, Evening Express,

Aberdeen, Evening Gazette, Middlesbrough, Evening News, Portsmouth, Evening News & Times,

Worcester, Evening Star, Suffolk,

Evening Telegraph & Post, Dundee, Evening Times, Glasgow Express & Star, Wolverhampton,

Furniture Record, Gas World, Glasgow Herald, Glasgow, Gravesend & Dartford Reporter,

The Guardian, Manchester, Guardian – Journal, Nottingham, Guildford & Godalming Times,

Halifax Daily Courier & Guardian, Harrow Observer and Gazette,

Herald & Express, Torquay, Hertfordshire Hemel Hempstead Gazette, Herts Advertiser

Huddersfield Daily Examiner, Ilford Recorder, Investors Guide, Lancashire Evening Post

Leamington Morning News, Leamington Morning Press, Leamington Spa Courier

Leicester Mercury, Lincolnshire Daily Echo, Liverpool Daily Post, Liverpool Echo

Lynn News & Advertiser, Kings Lynn, Machinery Market, Manchester Evening News

Maidenhead Advertiser, Melody Maker, Mercury and Herald, Northampton

Merthyr Express, Merthyr Tydfil, Glamorgan, Mid-Devon Times, Newton Abbot

Middlesex Advertiser & County Gazette, Newcastle Evening Chronicle

Northamptonshire Evening Telegraph, Kettering, Northern Daily Mail, West Hartlepool

Northern Echo, Darlington, Northern Whig, Nottingham Evening News, Nottingham Evening Post

Nuneaton Evening Tribune, Oldham Evening Chronicle & Standard, Oxford Mail, The Pauline

Policy Holder, Practical Wireless, Press and Journal, Aberdeen, Radio Times, Richmond Herald

Scarborough Evening News, The Scotsman, Scottish Daily Express, Scottish Daily Mail, Edinburgh

Sheffield Telegraph, Sidcup & Kentish Times, Sidcup, Southern Evening Echo, Southampton

South Devon Journal, Torquay, South Eastern Gazette, Maidstone, South Wales Echo, Cardiff

South Wales Echo & Evening Express, Cardiff, South Wales Evening Post, Swansea

The Star, Sheffield, Sunderland Echo, Co. Durham, Surrey Advertiser, Guildford

Telegraph & Argus, Bradford, Thames Valley Times, Time & Tide, Tonbridge Free Press, Kent

Torquay Times, Travel Trades Gazette, Warwick Advertiser, Weekly Scotsman

West Lancashire Evening Gazette, Western Daily Press & Bristol Mirror,

Western Evening Herald, Plymouth, Western Mail, Cardiff, Western Morning News,

Plymouth, Wolverhampton Express & Star, Yorkshire Post, Leeds, Yorkshire Evening News,

Yorkshire Evening Post, Leeds

Motoring Press

Auotcar, Car Hire, The Garage & Motor Agent, L' argus de l' automobile, Paris, Modern Transport,

Motorsport Magazine, The Motor, Motor Industry, Motor Trader, Motor Transport,

Motoring News, Renault Magazine

Ireland

The Belfast Newsletter, Belfast Telegraph, Belfast Times, The Cork Examiner, Evening Mail, Dublin

Evening Press, Dublin, Irish Independent, Irish Press, Irish Times, Dublin

International

Time Magazine (Atlantic Edition) Financial Post, Canada, World Press News.

Television and Radio Coverage

Information from Tellex Reports complied by Tellex Monitors Ltd.

Tonight (BBC TV) 21.2.61 (6.50pm) *Mac Hastings interviews Taxi driver, Jim Francis and Mchael Gotla*

News (ITV) 15.3.61 (5.55pm)

News (BBC L.P) 10.3.61 (6.30pm)

Ten O Clock (H.S) 3.4.61 (10.00pm) *Michael Gotla interviewed*

News (ITV) 26.5.61 (5.55pm)

News (H.S) 6.6.61 (6.00pm)

News (BBC TV) 6.6.61 (6.00pm)

Town and Around (BBC TV) 8.6.61 (6.40pm) *Christopher Jones interviews Jim Francis and Michael Gotla*

Newsreel (L.P) 8.6.61 (7pm) *Christopher Jones again interviews Jim Francis and Michael Gotla*

In the South East (H.S) 9.6.61 (6.15pm)

Any Questions (LP) 9.6.61 *Debate with Taxi driver and member of the public*

Ten O Clock (H.S) 10.00pm) *Erskine Childers interviews Mr J H Francis.*

This Week (ITV) 16.6.61 (8.00pm) *Desmond Wilcox interviews Michael Gotla*

News (ITV) 19.6.61 (5.55pm) *Tom St Barry interviews Michael Gotla*

News (L.P) 19.6.61 (6.30pm)

News (H.S) 19.6.61 (6.00pm) *In the South East 19.6.61 (6.15pm)*

News (ITV) 19.6.61 (7.00pm)

Town and Around (BBC TV) 19.6.61 (6.07pm)

Radio Newsreel (L.P.) 19.6.61 (7.00pm) *Report by Ray Colley*

News (BBC TV) 19.6.61 (8.30pm)

News (BBC TV) 19.6.61 (8.40pm)

News (ITV) 19.6.61 (9.25pm)

News (H.S) 19.6.61 (10.00pm)

News (BBC TV) 19.6.61 (10.10pm) *Report by Ray Colley*

Today's Papers (H.S) 20.6.61 (7.35am)

Today (H.S) 21.6.61 (7.15am) *Richard Anderson interviews both Taxi and Welbeck driver*

What the Papers Say (ITV) 22.6.61 (10.35pm)

News (ITV) 23.6.61 (5.55pm) *Neville Clark interviews Welbeck driver*

News Extra (BBC TV) 23.6.61 *Michael Gotla interviewed*

News (ITV) 26.6.61 (5.55pm)

Town and Around (BBC TV) 28.6.61(6.10pm) *Bernard Murphy interviews 2 Welbeck drivers & Commander Milner*

News (ITV) 28.6.61 (9.25pm) *John Whale interviews Mr Francis*

Town and Around (BBC TV) 29.6.61 (6.10pm) *Bernard Murphy interviews 3 Taxi drivers & Jim Buntin*

New From London (ITV) 29.9.61 (11.15pm) *John Whale interviews Sam Henderson (T&GW Union)*

News (BBC H.S) 10.7.61 (7.00am)

News (H.S) 10.7.61 (1pm)

News (BBC TV) 4.8.61 (6.00pm)

Town and Around (BBC TV) 4.8.61 (6.07pm)

News (BBC H.S) 4.8.61 (10.00pm)

News (ITV) 27.8.61 (6.05pm)

From the Local Press (BBC HS) 28.8.61 (6.45pm)

Changing Face of London (H.S) 4.10.61 (6.30pm) *Raoul Engel interviews Michael Gotla*

Michael Gotla being interviewed on television outside the
Welbeck Motors Crawford Street Showroom in London
in 1961. (Graham Walker – Welbeck Archive)